VIRUSES FROM SPACE AND RELATED MATTERS

Fred Hoyle
Chandra Wickramasinghe
and
John Watkins

UNIVERSITY COLLEGE CARDIFF PRESS
1986

First published in 1986 in Great Britain by
University College Cardiff Press
P.O. Box 78, Cardiff, CF1 1Xl
United Kingdom
and
51 Washington Street, Dover, New Hampshire 03820 USA

Copyright © 1986 Fred Hoyle, Chandra Wickramasinghe and John Watkins

All rights reserved. No part of this publication may be reproduced, stored in a retrieval system or transmitted, in any form or by any means, electronic, mechanical, photocopying, recording or otherwise, without the prior permission of the Publishers.

ISBN 0-906449-93-6

Printed in Great Britain by Alden Press, Oxford

Contents

List of Figures	v
Preface	viii
Acknowledgements	ix

Chapter

1. Viruses from Space and Related Matters (1985)	1
2. The Anatomy of a Conference Epidemic (1983)	39
3. The Plague of Athens – Was it Smallpox?	47
4. Influenza – A Genetic Virus with an External Trigger	57
5. Influenza in Some West-country Schools (1978)	65
6. Does Epidemic Disease Come from Space? (1977)	79
7. The Influenza Epidemic of 1977/78	87
8. Schools Influenza Survey 1985	105
Index	114

LIST OF FIGURES

Fig. 1　Histogram showing the distribution of influenza attack rate among independent schools in England and Wales during the 1977–78 pandemic.

Fig. 2　Whooping cough notifications in England and Wales from 1940 to 1982.

Fig. 3　Quarterly incidence of whooping cough, measles and infective jaundice in the City of Cardiff *vs* The Vale of Glamorgan.

Fig. 4　Percentage attack rates in households where one member succumbs to upper respiratory infection in the epidemics of 1968/69 and 1969/70 at Cirencester, England, according to data from Hope-Simpson.

Fig. 5　Percentage cases of acute upper respiratory infections in households where one member succumbs to influenza on day 0, from data collected by J.W. in 1984/85.

Fig. 6　Correlation of attack rates for Day pupils and Boarders in schools which had a mixture of both during the 1977/78 influenza pandemic in England and Wales.

Fig. 7　Deviations of attack rates of influenza about the mean attack rate for the at 25 school houses at Eton College during the 1978/79 pandemic. The deviations are computed relative to the standard deviation computed house by house.

Fig. 8　Correlation of influenza attack rates for Junior and Senior pupils (Junior, 5-13 years; Senior $\geq$ 14 years) during the 1978/79 pandemic.

Fig. 9　The orbits of known short-period comets lying mostly in the region between Mars and Jupiter.

Fig. 10　The attack rates of influenza in Prague and Cirencester (after Hope-Simpson).

Fig. 11   The spread of the Black Death through Europe.

Fig. 12   The fall-out of Rh-102 at various latitude intervals from the HARDTACK atmospheric nuclear bomb which exploded on 11th August 1958.

Fig. 13   Incidence of RS infections in England and Wales (CDR reports).

Fig. 14   Incidence of Influenza A in three geographically separate countries.

Fig. 15   Incidence of Parainfluenza type 3 (CDR reports).

Fig. 2.1   The absence record of 1257 children in the 6th grade in six Colombo schools (November and December 1982).

Fig. 2.2   The departure from the mean weekly dispensation rate of antibiotics for upper repiratory and ENT infections, expressed in units of the sample standard deviation.

Map 1   Map of South Wales showing location of schools examined.

Map 2   Map of Howell's School Llandaff showing 4 houses.

Fig. 5.1   Excess absences (percent) above normal for Howell's School. Three curves show absences for whole school, boarders only and first two forms only.

Fig. 5.2   Excess total absences (percent) above normal for Howell's, Llanishen High and St. Cyres.

Fig. 5.3   Excess total absences (percent) above normal for Cardiff High, Balfour House and Clifton College.

Fig. 5.4   Schematic plan of dormitories at Howell's for four Houses: Hazelwood, Taylor, Oaklands and Bryn Taff. Boxes represent rooms, partitions within boxes are occupied beds. Numbers show order of victims reporting to the school nurse. H stands for a week-day boarder who had influenza over the weekend.

Fig. 6.1 The infrared spectrum (right) of a polysaccharide 'dust' model compared (left) with that of the BN source of the Orion Nebula.

Fig. 6.2 Photograph of Comet Kohoutek 1973.

Preface

The first chapter of this book is an up-to-the-moment discussion of the concept that viruses from space are the primary cause of so-called infectious diseases. Often, after lecturing on this subject, we are asked the question: Where can we read about all this? General accounts were given some years ago by F. Hoyle and N.C. Wickramasinghe, in the books *Diseases from Space* (J.M. Dent, 1979) and *Space Travellers* (University College Cardiff Press, 1981). These accounts were based on more technical publications which are not now easily available. The opportunity has therefore been taken in the present volume to reprint a selection of these more technical publications, written by F. Hoyle and N.C. Wickramasinghe, unless otherwise stated.

> Fred Hoyle
> Chandra Wickramasinghe
> John Watkins

November 1985

Acknowledgements

Chapter 2 of this volume was published in *Fundamental Studies and the Future of Science* (ed C.W.) published by University College, Cardiff Press, 1984. Chapter 5 is based on an article originally published in *New Scientist* in 1978; Chapter 6 is based on a *New Scientist* article of 17 November 1977.

1

VIRUSES FROM SPACE AND RELATED MATTERS (1985)
(by F.H., N.C.W. and J.W.)

Introduction

It is generally thought that acute upper respiratory tract infections are caused by the intake of viral particles that were previously exuded by some other person, or in rare cases by some other animal[1,2]. Yet little or no evidence capable of standing up to critical analysis has ever been presented in support of this widespread opinion, which appears to have arisen through historical accident rather than through accurate observation and experiment. Following Pasteur's classic experiments on alcoholic fermentation and silkworm diseases it became established that some human diseases arise from the transmission of bacteria from person-to-person, and since in the later decades of the 19th century there was no appreciation of the difference between viral and bacterial diseases the concept of infection by person-to-person transmission became applied to all diseases, a point of view that appeared to gain support when in 1892 Pfeiffer[3] mistakenly implicated the bacillus *H.Influenzae* as the causative agent of influenza. A few epidemiologists, notably Charles Creighton in Britain, continued to protest that the evidence contradicted the rising tide of medical opinion, but in an age when few students had the leisure, affluence and inclination to examine the facts for themselves, the 19th century belief became set rigid in the educational system.

Once a false belief becomes established it is very difficult to get it out, essentially because the system invents supposed facts in order to support it. Two of the present authors[4] had an experience of this process in action following the peculiar epidemic outbreaks of influenza A in the winter of 1977-78, peculiar because of the return of influenza subtype H1N1 with a variant dating apparently from the year 1950. It is commonly stated that the person-to-person transmissibility of influenza is proved by very high attack rates in institutions such as barracks and boarding schools. Yet a survey of boarding schools in 1977–78 involving a total of more than twenty thousand pupils with a number of victims estimated to be some 8880 for an average attack rate of about 30%, yielded the distribution of attack rates shown in the histogram of Fig.1. In fact, only three schools out of more than a hundred at the extreme upper end of an approximately exponential distribution had the very high attack rates which have been claimed to be the norm.

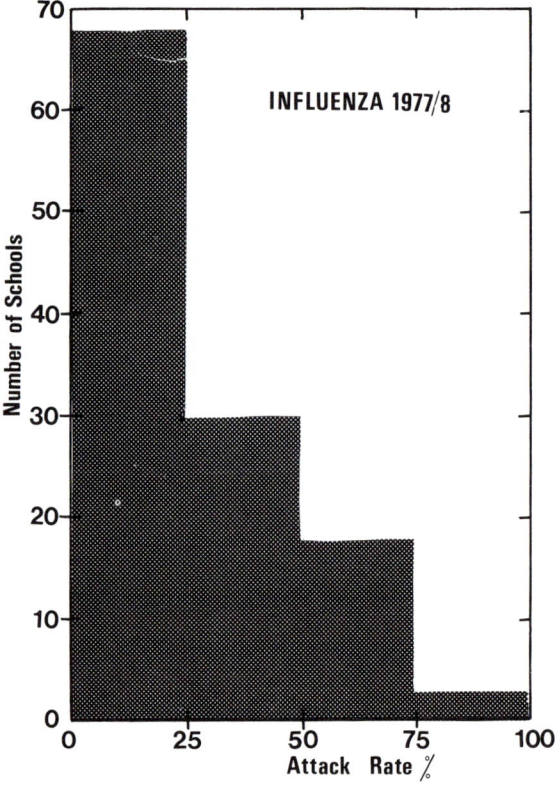

Figure 1.
Histogram showing distribution of influenza attack rate among independent schools in England and Wales during the 1977–78 pandemic.

All the diagnoses involved in these data were made by school medical staffs in advance of our enquiries. Possibly other respiratory infections became associated with influenza in the diagnoses, but since January and February 1978 were months of influenza epidemics, and since children of school age had no established immunity against influenza H1N1, the bulk of the reported cases were very likely of this disease. And even if, in the absence of isolates or serological tests, one were sceptical of explicit diagnoses, the cases were certainly of acute upper respiratory infection, to which just the same remarks and conclusions would apply regardless of the explicit viruses involved. The schools in question were fee-paying, all with boarders sleeping together in dormitories. The degrees of association of pupils in dormitories, classes and at meal times were not much different from one school to another, and if the virus or viruses responsible for the 8880 cases were passed from pupil to pupil, much more uniformity of behaviour would have been expected. Already in Fig. 1 we have evidence of great diversity, with a hint that the attack rate experienced by an individual school depended on where it was located, with some schools being in fortunate places and some in unfortunate places.

The alternative to the person-to-person transmission of a virus is that it falls from the air. For semantic convenience we refer to falling from the air as vertical incidence and to person-to-person transmission as horizontal transmission. Although in this article we are concerned to argue the case for vertical incidence as the cause of most acute upper respiratory infections, it is to be emphasised that we are not making this claim for all viral diseases. While we think that all viral diseases arise in the first place by vertical incidence, it is possible for a virus to establish a reservoir in the human population such that the chance p of contracting the associated disease by human contact is greater than the chance q of contracting the disease through vertical incidence. Normalising so that $p+q=1$, there are the possibilities $p >> q$, $p \simeq \simeq 1$; $p \simeq q \simeq \frac{1}{2}$; $p << q$, $q \simeq 1$. Diseases in the first of these categories are truly infectious and can be moderated greatly through the old-fashioned method of isolating victims. Indeed one could say that it is just those diseases, as for example smallpox, which the medical profession found to be successfully treated by isolation, that constitute the truly infectious category $p \gg q$, $p \simeq 1$. In this article we are concerned with the opposite more numerous category, $p \ll q$, $p \simeq 1$, which includes most acute upper respiratory tract illnesses. Data for measles, the discussion of which goes beyond the scope of this article, suggests that measles belongs to the intermediate category $p \simeq q \simeq \frac{1}{2}$. We suspect it is the intermediate nature of measles which explains why the medical profession is divided in its opinions on whether the isolation of victims would or would not effectively stamp out the disease. If our assignment of measles is correct, isolation would appreciably reduce the number of cases but would not stamp out the disease. To stamp out a disease q must be strictly zero, requiring that the input of a virus to the Earth's upper atmosphere shall have ceased.

As regards the input of viruses to the Earth's atmosphere, the particles responsible for the strong ultraviolet component of the zodiacal light must have radii of order 30 nm, the scale of viruses. The density of such particles necessary to explain the observed strength of the zodiacal ultraviolet is remarkably high, implying an addition $\sim 10^4$ tons per year to the Earth's atmosphere, a total $\sim 10^{26}$ particles added annually. This number may be compared with an epidemic of disease in which each of $\sim 10^9$ humans sheds $\sim 10^{11}$ viral particles, for a total shedding of $\sim 10^{20}$ particles. If only a small proportion of the small zodiacal particles are viruses, if only a small proportion maintain viability, and if only a small proportion interact pathogenically with terrestrial plants and animals, the incident number would nevertheless be so vast that it could dominate horizontal transmission, even under extreme epidemic conditions.

It is commonly assumed that viral diseases are caused by the input to a victim of particles that are substantially identical to the output of viruses from the victim. This assumption is normally made for horizontal transmission, but it is not necessary for vertical incidence. All our data and all our arguments require a causative agent or trigger to fall from the air, but the resulting disease could be caused by the association of the causative agent with dormant viral particles present already in the victim. Or the whole virus could be involved as with horizontal transmission. The evidence we shall present does not distinguish these possibilities.

Bacterial diseases can also be thought of in terms of the categories $p \gg q, p \simeq 1; p \simeq q \simeq \frac{1}{2}; p \ll q \simeq 1$, but with the last category less common than for viral diseases, i.e. dominant vertical incidence being less common. One bacterial disease that is difficult to explain except by vertical incidence, however, is whooping cough. Pertussis has for long been known to occur in cycles of about 3.5 years, which used to be explained on the density of susceptibles theory, namely that after children susceptible to the disease become exhausted by a particular epidemic it was then supposed to take about three and a half years for new births to rebuild the density of susceptibles to the level at which a further epidemic would run. Thus the periodicity of this theory should have been a function of population density, with the shorter periods being found in inner city areas of very high density and either long periods or no periodicity at all in lightly populated country areas. But the periodicity was found to be everywhere the same, in town and country alike, and from one country to another. Figure 2 shows the record of notifications for England and Wales over the period 1940-82. If the theory had been correct, the sudden reduction in the density of susceptibles brought about in the 1950s by the introduction of an effective vaccine should have greatly disturbed the periodicity, or even destroyed it altogether. Yet the periodicity persisted exactly as before, but with the total number of cases much reduced.

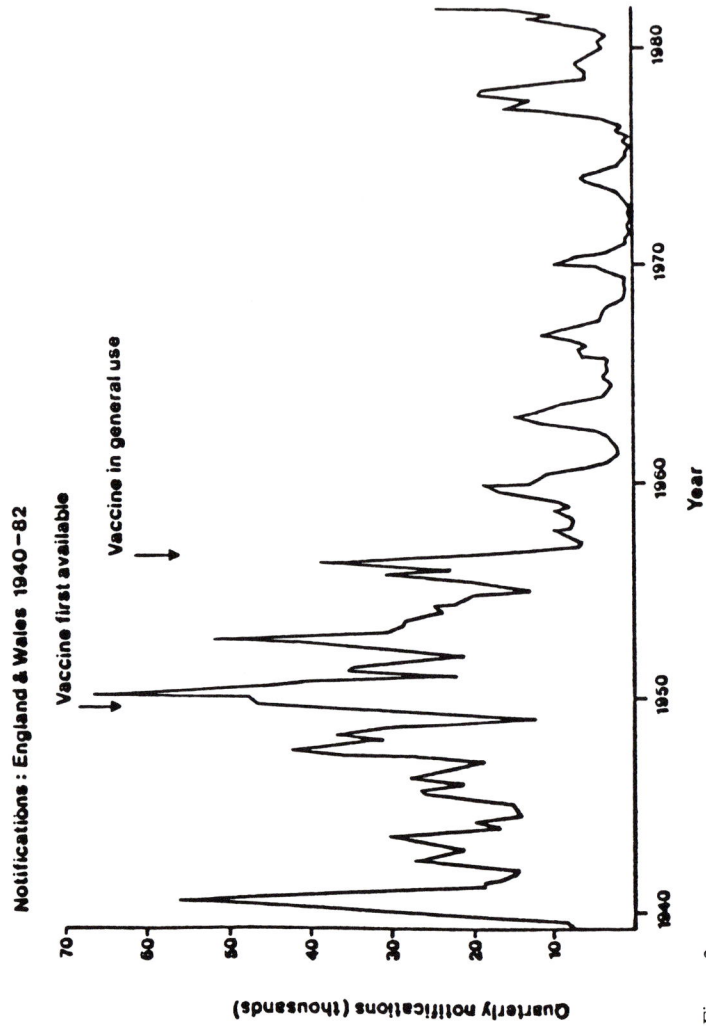

Figure 2.
Whooping cough notifications in England and Wales from 1940–1982.

General Evidence Against the Horizontal Transmission of Certain Diseases

If infectious diseases were propagated from person-to-person according to the commonly-held view then people living in high-density city areas should be significantly more subject to disease than people living in lightly-populated areas. From normalised attack rates plotted as a function of population density it would be possible therefore to prove the correctness or otherwise of this point of view. The circumstance that such data do not appear to exist, despite the cogency they would have, is interesting psychologically. Whereas people are avid to collect the slightest scraps of information that support conformist opinion, they are unremitting in their determination not to collect, or even to notice when collected, data which prove the opposite. It really needs no more than the absence of this simple but critical information to see that the commonly-held view must be wrong.

One can say in general terms that if any major discontinuity existed between town and country the population at large would easily be aware of it. Also in general terms, one of the present authors has shifted on occasions from general practice among a major city population to standing locum in highly rural areas, without any difference in morbidity patterns being qualitatively apparent. Thousands of general practitioners must have experienced similar comparisons without any discontinuity of pattern being emphasised and reported. On a more quantitative level, Figure 3 shows data collected by Dr. P. Jenkins, the Community Health Officer for the City of Cardiff. It gives data covering the three diseases of so-called infective jaundice, whooping cough and measles, obtained quarterly from the heavily-populated Cardiff city area (after normalising to 100,000 population) and from the Vale of Glamorgan, much of which is very rural. Thus each disease in each quarter of a period of three years yields a point in Fig. 3. This is except for measles which was so prevalent in one particular quarter that the corresponding point for that quarter could not be plotted without prejudicing the scale of

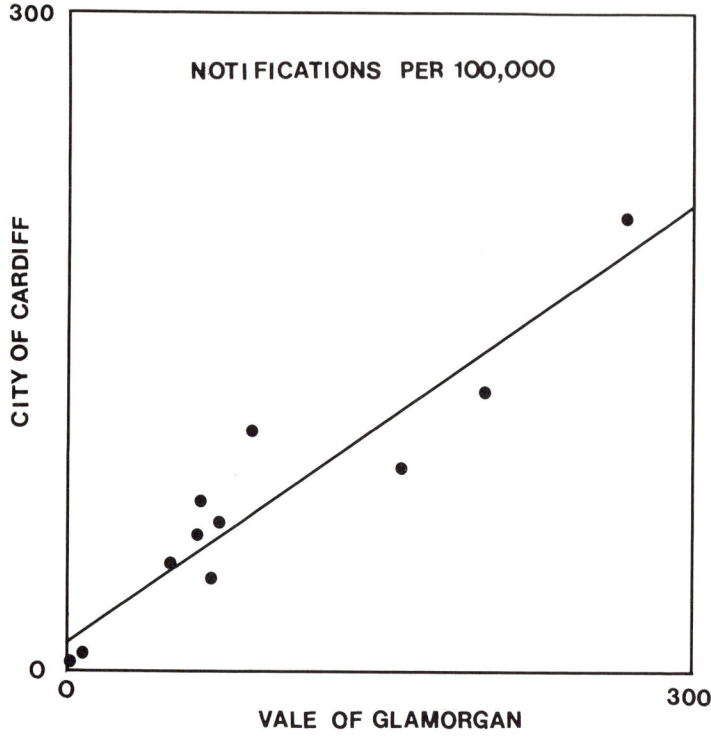

Figure 3.
Quarterly incidence of whooping cough, measles and infective jaundice in the City of Cardiff *vs* The Vale of Glamorgan.

the figure. The one missing point lies on the line defined by the other eleven points, but far away to the right of the figure. Such bias as one can see in Fig. 3 goes the wrong way for horizontal transmission. It is the lightly populated Vale of Glamorgan that on a normalised basis appears worse affected. Of course one can always invent the hypothesis that standards of reporting are higher in country practices than in the cities, but one of us as a general practitioner in a city area would naturally dispute this suggestion. At all events, general experience, together with the data of Fig. 3, suggests that there is no marked difference between town and country, as one would expect for vertical incidence but not as one would expect for horizontal transmission.

Ockham's razor warns us against inventing a 'multiplicity of hypotheses', a warning which some have seen fit to interpret as an edict proscribing the consideration of new ideas. What the warning really means is that we should be on our guard against the invention of a multiplicity of unsubstantiated hypotheses in order to defend conformist views against awkward facts. For example, it is in our opinion an *ad hoc* hypothesis to suppose that city populations possess greater immunity against infectious diseases than rural populations, and to such an extent that the greater exposure which city populations experience with respect to person-to-person transmission is almost precisely compensated by their greater immunity. A similarly *ad hoc* hypothesis would also be required to explain why individuals whose occupations involve exceptional hazards with respect to person-to-person transmission, for instance dentists and cashiers in banks, newsagents and large stores, nevertheless have records of upper respiratory infections that are not noticeably abnormal.

Looking over case notes dating from 1970 in a general practice, one of us (J.W.) identified 16 pairs of twins with ages between 6 months and 14 years. Of the 118 instances in which one twin was consulted for acute upper respiratory infection the corresponding twin succumbed to a similar infection in 28 instances. Since twins in the age range in question are found almost perpetually together, the opportunity for person-to-person transmission would be maximal in these twin-to-twin relationships. Yet in only 24 percent of instances did the second twin become a victim, which is not an impressive fraction, particularly as attack rates during epidemics of upper respiratory infections tend to run typically at about the 25 percent level among the general population. An epidemic will not run according to horizontal transmission unless each victim infects at least one other victim, thereby establishing a supercritical chain relation. A transmission probability as low as 0.24, which this data for twins yields on the horizontal transmission hypothesis, would therefore be quite insufficient to establish a

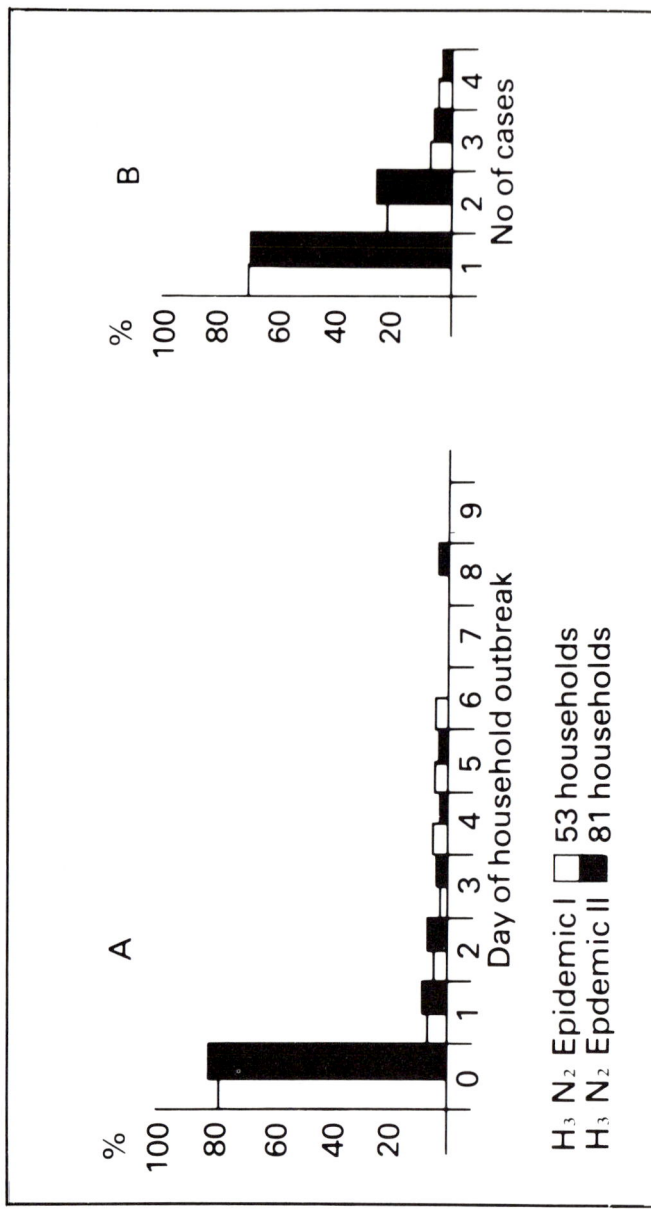

Figure 4.
Percentage attack rates in households where one member succumbs to influenza in the epidemics of 1968/69 and 1969/70 at Cirencester, England according to data from Hope-Simpson.

supercritical chain. Only if a stricken twin contrived to infect others much more readily than his or her own twin could adequate transmission be attained.

Following in the steps of Charles Creighton, Edgar Hope-Simpson was the first person in recent years to bring the hypothesis of person-to-person transmission rigorously under the hammer. Hope-Simpson had the idea of defining a set of households by the condition that one member succumbs initially to Influenza A. He then observed the subsequent fates of other members of the households thus defined, finding them to develop no greater proportion of attacks than would be expected for the population at large.[5,6] Figure 4 gives Hope-Simpson's results for epidemics of H3N2 in 1968/69 and 1969/70, shown in the histograms as I and II respectively. Besides the fraction of subsequent cases being normal for the population at large, no well-defined subsidiary peak occurred 12 days after the first cases were reported, as would be expected from incubation if horizontal transmission had been occurring. Hope-Simpson's results have been fully confirmed by Mann *et al.*,[8] and by Philip *et al.*[7]

The unusual circumstances in 1984-85 that few true cases of Influenza A were reported anywhere in the world up to mid-February 1985 (INFLU Centre, London[9]) permitted the behaviour of other sources of upper respiratory infections to be examined in a manner similar to that used by Hope-Simpson. Starting in May 1984, a total of 80 households were defined, again by the criterion that one member presented themselves with an acute upper respiratory infection, and then subsequent histories of other household members were studied by one of us (J.W.), with the results shown in Fig. 5. The interesting point emerges that upper respiratory infections quite generally are like influenza, without evidence of person-to-person transmission, which if it had occurred would have caused an incubation peak of cases to occur two or three days after the initial attacks on day zero.

A proportion of the fee-paying schools in the survey already mentioned had both day pupils and boarders. The boarders were exposed to close person-to-person contacts for 24 hours a

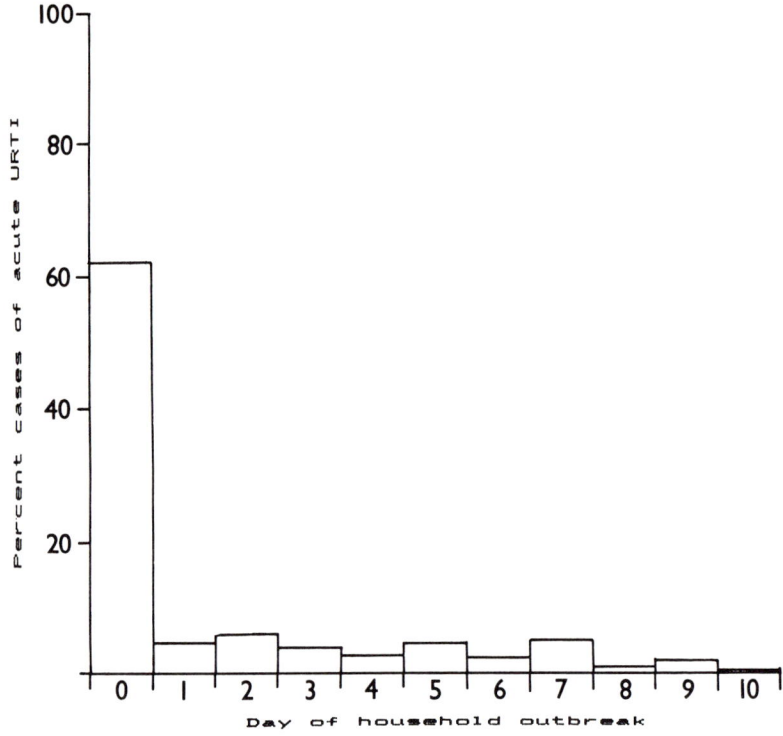

Figure 5.
Percentage cases of acute upper respiratory infections in households where one member succumbs to influenza on day 0, from data collected by J.W. in 1984/85.

day, whereas the day pupils were only some 8 hours at school, with the remaining 16 hours spent under non-institutional conditions, conditions having fewer person-to-person contacts generally. If there were any substance to the claim of high attack rates in institutions, used to bolster the person-to-person transmission hypothesis, the overall attack rates on boarders should have been significantly higher than it was on the day pupils. With each of the schools in question represented by a point in Fig. 6, the results gave essentially a scatter diagram. Whatever slight bias there is about the 45° line in this diagram disappears for a line of slope 40°, and this is within the expected statistical fluctuation. There are many instances in which the day pupils experienced considerably higher attack rates than boarders, a situation that defies the imagination to explain according to person-to-person transmission, for we would have to suppose that after leaving school the day pupils encountered more seriously infective contacts than were present at school, and that they did so *systematically* in order to explain high attack rates above 70 percent for the day pupils in some of the cases.

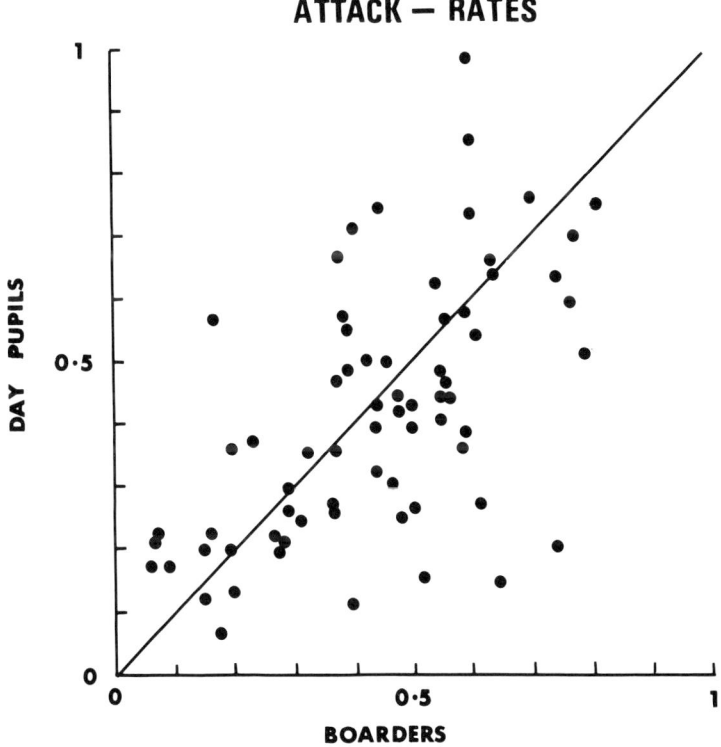

Figure 6.
Correlation of attack rates for Day pupils and Boarders for schools which had a mixture of both during the 1977/78 influenza pandemic in England and Wales.

In the next section, we shall see that vertical incidence is expected to lead to intricate patchy details in attack rates, with some localities relatively safe from attack and others relatively dangerous. On this view, schools that happened to be in relatively safe areas would have their boarders staying comparatively safe the whole time, whereas the day pupils would go out from the school into comparatively dangerous areas and so would experience significantly higher attack rates. And of course the opposite situation would occur for other schools, thereby producing the scatter shown in Fig. 6.

Attack rates of around 30 percent were found most useful for studying variations within school boundaries, since very high attack rates evidently preclude variations being found, while low attack rates gave inadequate statistical weight. Eton College had 441 victims among 1248 pupils, for an attack rate of 35 percent, with high statistical weight because of the large number of pupils. We were fortunate that Dr. J. Briscoe, the Medical Officer at Eton, had for long been puzzled to understand how his observations could be explained in terms of pupil-to-pupil transmission. Consequently, Dr. Briscoe had collected comprehensive information giving the distribution of victims in some 25 school houses. The houses averaged about 50 pupils each, with about 17 cases expected as the mean number of victims. Such numbers were very suitable for computing standards deviations, with the results shown in Fig. 7. Two houses had excess morbidities of 4 standard deviations, two had deficits of about 4 standard deviations, while one house (COLL) had a remarkable deficit of 6 standard deviations. Since pupils in the different houses were mixed in classes and at games these enormous fluctuations from a random distribution are quite inexplicable it seems to us in terms of horizontal transmission. The Eton College results imply that the school was hit vertically by the influenza virus during the night hours, or possibly at a weekend, and that the vertical incidence was patchy enough to distinguish between the locations of the various houses, some houses happening to lie in safe areas and others in dangerous areas. Dr. Briscoe informed us that similar effects had occurred in other influenza

VIRUSES FROM SPACE AND RELATED MATTERS (1985)

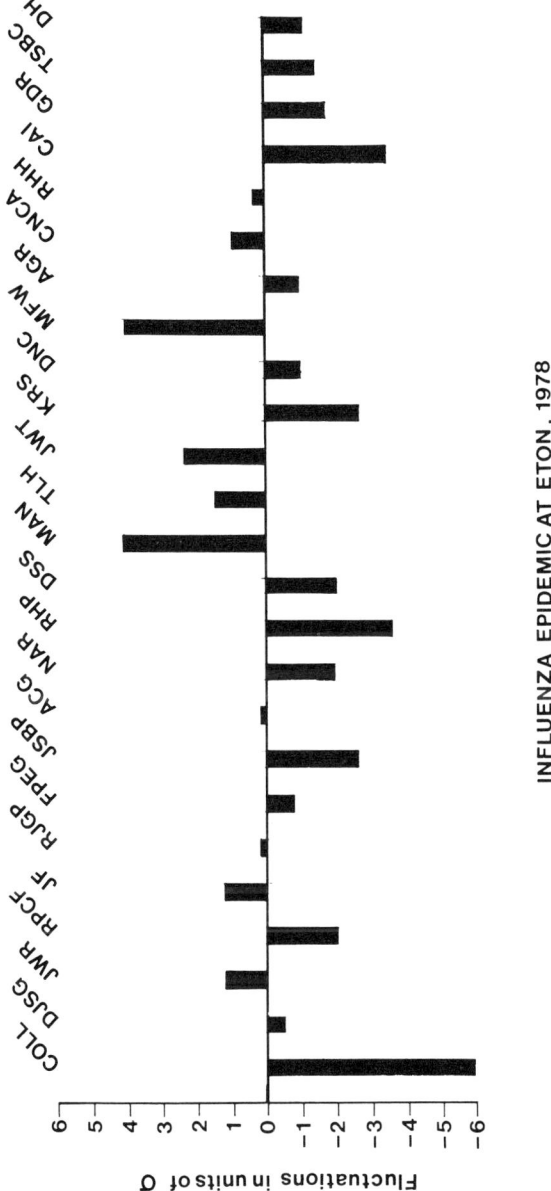

Figure 7.
Deviations of attack rates of influenza above the mean attack rate for the 25 school houses at Eton College during the 1978/79 pandemic. The deviations are relative to the standard deviation computed house by house.

epidemics, with the identities of the lucky and unlucky houses being different from the situation in 1978. A patchy vertical incidence would of course not be reproducable in its details from one epidemic to another, so this too would accord with the vertical incidence hypothesis.

We end this section with a somewhat different issue. Younger children have sometimes been found to be more susceptable to influenza than older children. Usually it is not possible to distinguish how far the greater resistance of older children is inherent and how far due to already established immunities. Since no children of school age in 1978 had any established immunity to the H1N1 subtype, and since some schools in our survey had both junior and senior pupils, it was possible to compare attack rates that gave information largely free of the immunity factor. Results are shown in Fig. 8 where each point refers to a school having both junior and senior pupils. These data suggest that inherent resistance has little to do with age, implying that differences observed in other years were related to the immunity factor. Such asymmetry as one sees in Fig. 8 about the 45° line (after noting the two very low points marked heavily to catch the eye) would be removed by increasing the 45° slope to a slope of 50°. A slight bias in this sense could have arisen from a minority of cases where the virus subtype was still H3N2 rather than H1N1, with older children having better immunity to H3N2.

The Vertical Incidence Theory

According to medieval lore, diseases come from comets, and according to our view this is true, but only in a broad sense. We cannot maintain the dramatic position that ferocious new diseases come from spectacular comets, because for every spectacular comet there are almost certainly very many smaller ones. The smaller comets may not only evaporate more material collectively than large ones but the effect of the Earth crossing almost precisely the track of a small comet would lead to a greater addition of evaporated particles to the terrestrial atmosphere than would a more distant relationship to a large comet, as for instance a distant relationship to Halley's comet.

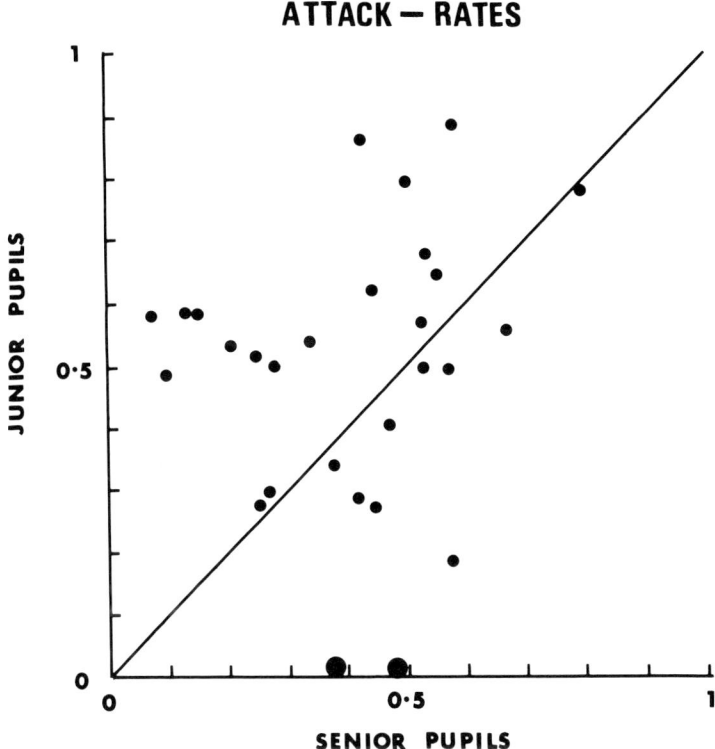

Figure 8.
Correlation of influenza attack rates for Junior and Senior pupils (Junior, 5–13 years; Senior ≥ 14 years) during the 1978/79 pandemic.

Support for this position comes from an analysis by Z. Sekanina[10] of some 20,000 orbits of meteors, meteors being small particles typically with sizes ~ 0.1 mm also evaporated from comets. A minority of the orbits could be associated with known comets but the majority could not. It is perhaps possible to understand both the origin and the demise of the medieval lore in these terms. Over a time scale of historic length there must have been special situations with the Earth in close proximity to a large comet. If following such special situations serious attacks of disease occurred, particularly if entirely new diseases appeared, their association with the comets would be an obvious and natural deduction. But then as other less close comets appeared in the sky, and were not followed by spectacular outbreaks of disease, the belief would be thrown first into doubt and then into ridicule.

Figure 9 gives a rather mild idea of the complexity of the situation. It shows only the orbits of the so-called Jupiter family of comets. One has to imagine thousands upon thousands of orbits of smaller comets added to Fig. 9 in order to appreciate the real position. At all events, we can see that the Earth is perpetually embedded in a halo of evaporated cometary material, some of the material newly evaporated, without much in the way of exposure to solar ultraviolet light. Following the development of the panspermia theory by Svante Arrhenius[11] early in the present century, it became fashionable to discount the theory by claiming that microorganisms in space would be destroyed by UV, by X-rays, by low pressure, by temperature etc. The circumstance that all these claims have turned out to be untrue is indicative of the correctness of the spaceborne theory. All that was wrong with the panspermia theory was that it did not go nearly far enough. While data on the radiation hardiness of viruses is less complete than for other kinds of micro-organism, on the general argument that simpler systems are usually more hardy than complex ones, viruses can be expected to withstand conditions in space at least as well as bacteria and algae. The following is a quotation from R.B. Hoover, F. Hoyle, N.C. Wickramasinghe, M.J. Hoover, and S. Al-Mufti:[12]

"Fowler *et al.*, (*Nucleonics, 18*, 1960, 102) reported a species of *Pseudonomas* living in a nuclear research reactor where the accumulated dosage was estimated at more than a million rads. *Micrococcus radiodurans* can also survive exposures of megarads. Nassim and Jones (in *Microbial Life in Extreme Environments*, ed. D.J. Kushner, Academic Press 1978) report the example of an exposure estimated to have caused of the order of 10,000 breaks in the DNA of these bacteria. Yet the bacteria repaired this immense damage by an intricate process

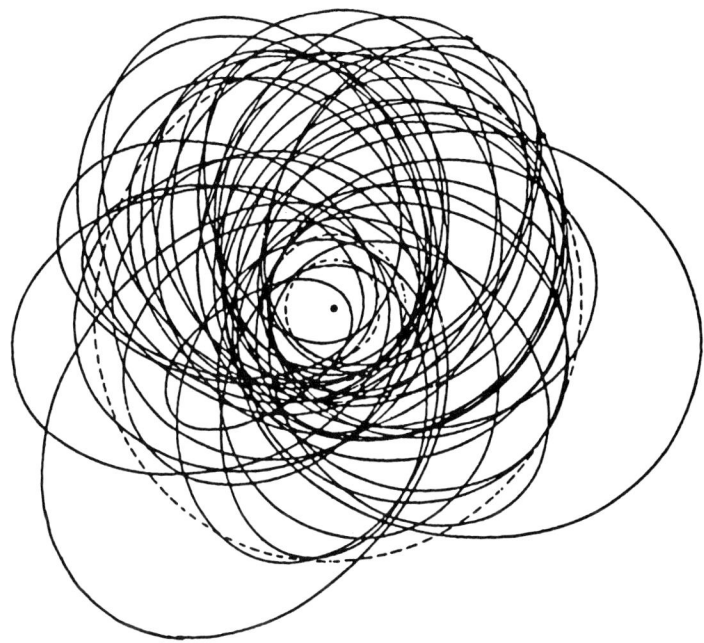

Figure 9.
The orbits of known short-period comets lying mostly in the region between Mars and Jupiter.

of snipping and inverse base-copying. Although viruses could not perform such a feat autonomously, they appear to be able to use the enzymic equipment of host cells to repair great damage and even to cannibalise bits from several otherwise defunct particles into a single active particle.

"It has been established that many diatom species are capable of thriving in environments containing extremely high concentrations of unusually lethal radio-isotopes such as plutonium, americium, strontium, etc. Diatoms (which are algae) thrive in highly radioactive waste ponds, including the U-pond and Z-trench containing over 8 kg of various radio-isotopes of plutonium at the Hanford facility. Not only do diatoms live in this environment but they seem to have a remarkable affinity for plutonium. Emery *et al.*, (Report prepared for the U.S. Atomic Energy Commission under contract AT(45-1) 1830; BNWL-1867, p.44, 1974) state that the algae of these ponds concentrated americium-241 three million-fold, while certain isotopes of plutonium were accumulated to 400 million times their concentration in the surrounding water. In this environment diatoms grew in great abundance while continuously subjected to high levels of X-rays, gamma-rays, alpha and beta particles.

"Hagen *et al.*, (*Space Life, 3*, 1971, 108) investigated the effect of temperature and pressure on the survival of microorganisms. In the NASA Jet Propulsion Laboratory Space Molecular Facility bacteria and bacterial spores were exposed to a simulated deep space environment. The organisms included *Bacillus subtilis var.niger, Staphylococcus epidermidis* and a species of *Micrococcus* isolated from Apollo 11 before launch. The specimens were subjected to a hard vacuum $\sim 10^{-10}$ atm and to temperatures ranging from $-124°C$ to $+59°C$. These authors state: 'Bacterial survival was better in the test environment at all temperatures, than cells held at ambient room conditions'. Such results clearly show that the effects of hard vacuum and low temperature such as are encountered in deep space are not lethal to these microorganisms. On the contrary, it seems greatly to improve their survivability compared to conditions at the surface of the Earth.

"The question of the survivability of terrestrial type microorganisms in alien environments is now new. Seckbach and Libby (*Space Life, 2*, 1970, 121) exposed algae to conditions that simulated the atmosphere of Venus, CO_2 at 50% and a pressure of 50 atm in acid. The green alga *Scenedesaus sp.* produced larger cells and showed higher activity in the simulated environment than in the laboratory control. *Cyanidium caldarum*, a thermophilic/acidophilic alga collected from the acid sulphate springs at Yellowstone National Park thrived greatly in the simulated Cytherean atmosphere. It also produced larger cells than the control."

The discovery of such properties reverses sharply the logic used formerly against the panspermia theory. Instead of the problem being survival in a space environment, a seemingly insoluble problem now confronts conventional theory, namely to understand how micro-organisms that are supposed never to have been outside the terrestrial environment could have acquired properties so remarkable and so profoundly suited to a space environment. Because the Earth and the finely-divided cometary material in the halo around the Earth are not comoving, the high terrestrial atmosphere is essential if micro-organisms are to make a soft landing here. Speeds relative to the Earth are so high that micro-organisms would be destroyed by hitting a hard surface, for instance the surface of the Moon. The maximum size for the safe entry of biological material is ~ 0.1 mm. This is under the most favourable conditions of speed and geometry. For smaller sizes, those of bacteria and viruses, the situation is not restrictive, however. Particles of the sizes of bacteria and viruses, indeed particles not larger than 0.01 mm, land 'soft' in the high atmosphere, permitting them to retain viability, and subsequently to fall gently downwards as active agents.

The Descent of Small Particles Through the Terrestrial Atmosphere

The lower atmosphere, the troposphere, has a height which varies from about 18 km in the tropics to 10 km in temperate latitudes to 7 km in polar regions. Small particles over the whole size range from viruses with diameters ~ 100 nm up to colonies of bacteria descend comparatively quickly from the top of the troposphere to ground-level. The troposphere is a region of falling temperature with increasing height, a physical condition permitting vertical air movements to occur readily, thereby causing water vapour to be carried upwards from the surface regions to essentially the top of the troposphere. The falling temperature with increasing height makes the water vapour supersaturated, but the temperature is not usually so low that the supersaturated water vapour condenses spontaneously into ice crystals. Initially-existing nuclei are required for the water to condense around, and the small particles in question provide such nuclei. Thus small particles on reaching the troposphere from above become condensation centres around which much larger ice crystals form. Because they are less impeded by air resistance than the original small particles themselves, such ice crystals fall with much increased speeds. As they descend into warmer air the ice crystals usually become melted and the resulting water droplet may either fall to the ground as rain, or it may become partially evaporated and as a smaller droplet remain suspended in the air. Normal precipitation rates are such that this process, often involving repeated cycles of condensation and evaporation of the ice crystals and water drops, serves to wash-out the troposphere of small particles in a time-scale of a few weeks. Exceptional conditions may extend the time-scale, however, for example in desert regions or as in the unusually cold weather experienced in northern latitudes during January and February 1985.

Above the top of the troposphere, the tropopause, the temperature undergoes inversion through the stratosphere, rising from $\sim -55°C$ at the tropopause to a maximum of $\sim 3°C$ at a height of ~ 50 km. The physical reason for this temperature inversion is that the ozone in the height range in question, the stratosphere, absorbs solar ultraviolet light very strongly shortward of 3000 Å, thereby giving an energy input into the stratosphere. The dynamical effect of the temperature inversion is to inhibit greatly the generally free movement of air such as occurs in the troposphere. Travellers by air will be familiar with the difference between the clarity of the lower stratosphere into which airplanes normally climb and the cloudy turbulent troposphere below. Free air movement in the stratosphere is limited to west-to-east movements along parallels of latitude, of which the most violent are the jet streams. The effect of the free west-to-east movement is to produce a general uniformity with respect to longitude in the stratospheric distribution of small atmospheric particles. If the Earth were smooth at its surface, we would therefore expect any incoming pathogens from space to arrive at ground level at more or less the same times along a given parallel of latitude (although local weather patterns could still introduce modest fluctuations). But because the troposphere has a marked dependence of height on latitude, we would not expect different latitudes to behave similarly, unless the particles happened to be so large that they were able to fall rapidly without much regard to air resistance.

Since the surface of the Earth is actually not smooth, in particular the Himalayas project about halfway up to the stratosphere, the rule of contemporaneous incidence along parallels of latitude is likely to be inappropriate in regions of high relief. Nevertheless, with the exception of the Rocky Mountains of North America there is a belt around the Earth from about 45°N to 60°N where the land is not much above sea-level and where the rule should be applicable. Prague, the

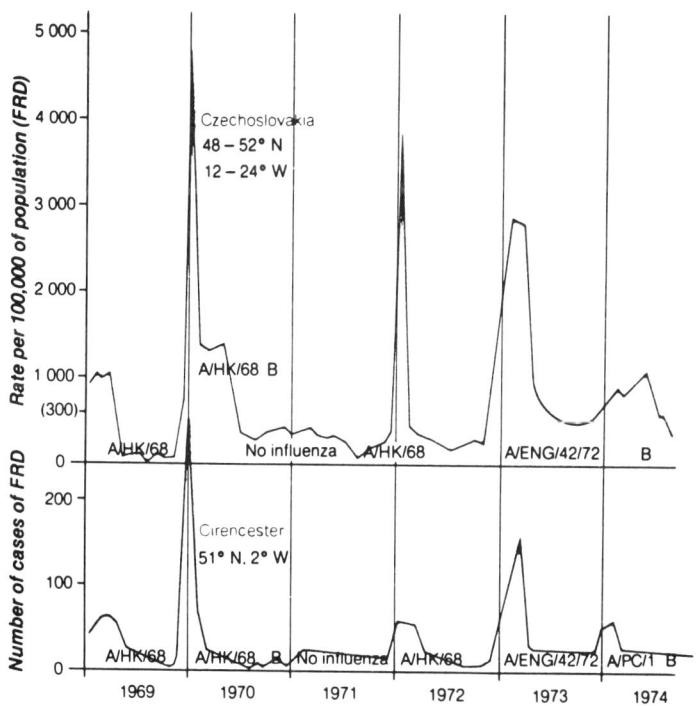

Figure 10.
The attack rates of influenza in Prague and Cirencester after Hope-Simpson.

capital city of Czechoslavakia lies a little north of 50° and Cirencester in England has a similar latitude within about 1°. Hope-Simpson has noted the similarity shown in Fig. 10 between his influenza records for the Cirencester district and the Czech records.[13]

A particle of radius 10 μm falls through the lowest ten kilometres of the stratosphere at a speed ~ 1 cm s^{-1} and thus takes only a few days to cover what for smaller particles is the slowest part of their downward journey. All particles fall comparatively rapidly through the upper atmosphere above the stratosphere, and then more and more slowly down through the stratosphere. A particle with the size of a typical bacterium, ~ 1 μm, falls through the lowest ten kilometres of the stratosphere at a speed of about 2.10^{-2} cm s^{-1} and thus falls in a time-scale of $\sim 5.10^7$ s, i.e. about 2 years. Because there is more of the stratosphere through which such a particle must fall in high latitudes than in the tropics (recalling that the tropopause is higher in the tropics) the slow part of the journey is more extended the higher the latitude. A bacterium falling in ~ 1 year in the tropics would fall in ~ 2 years in temperate latitudes and in 2 to 3 years towards the poles, a situation that is broadly consistent with historical records of the dates of outbreak of the Black Death in various latitudes, as shown in Fig. 11. It is interesting to notice the perturbation of contours of contemporaneous outbreak produced by the Alps, which mountains rise to about half the height of the tropopause.

If a particle of the size of a typical virus, a particle say with diameter ~ 0.1 μm, fell under gravity through still air the time-scale for the slowest part of the journey through the bottom ten kilometres of the stratosphere would be $\sim 10^9$ s, i.e. about 30 years. This is so slow that other means of descent involving large scale air movements in the stratosphere have to be considered. Although vertical mass movements of air are feeble compared to those in the troposphere, some vertical stratospheric movement takes place despite the inhibiting effect of the inverted temperature gradient. The physical cause of mass stratospheric movements is the equator to pole temperature difference which is available to work a heat engine crossing

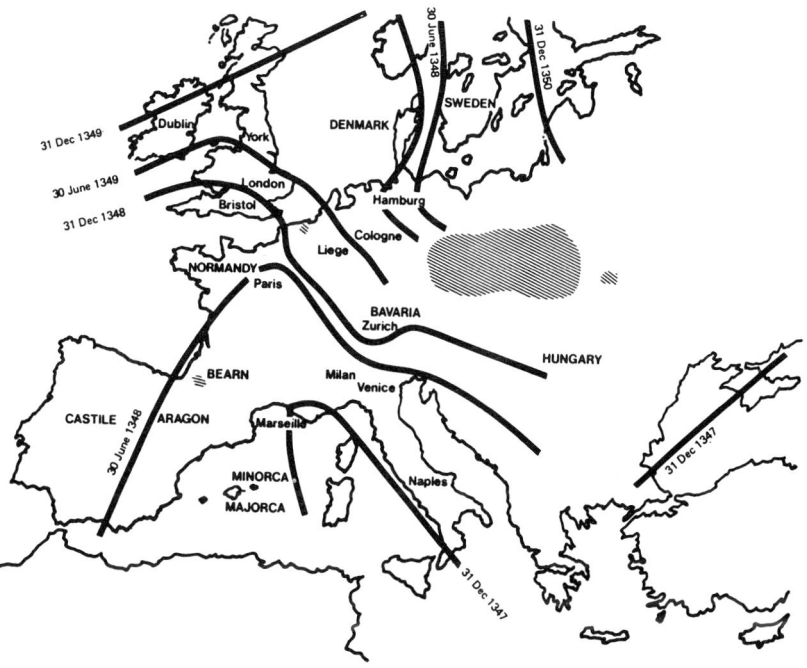

Figure 11.
The spread of the Black Death through Europe.

parallels of latitude, a heat engine that operates more strongly the larger the temperature difference – i.e. much more strongly in winter than in summer. A similar consideration applies also in the troposphere, where an engine crossing parallels of latitude transfers heat from tropical regions towards the poles, again more in winter than in summer. The heat engine in the troposphere is that which we experience in cyclonic storms.

Ozone measurements can be used to trace the mass movements of air in the stratosphere. Such measurements show a winter downdraft that is strongest over the latitude range from 40° to 60°. Taking advantage of this annual downdraft, individual viral particles incident on the atmosphere from space would therefore reach ground-level generally in temperate latitudes, which therefore emerge from these considerations as the regions of the Earth where upper respiratory infections are likely to be most prevalent, once again on the supposition that the Earth is smooth. The exceptionally high mountains of the Himalayas, rearing up through most of the height range to the stratosphere, introduce a large perturbation on the smooth condition, which may be expected to affect adversely this particular region of the Earth, especially regions lying downwind of the Himalayas, particularly China and S.E. Asia. In effect, the Himalayas are so high that they could act as a drain plug for most of the viruses incident on the atmosphere at latitude $\sim 30°N$, the large population of China being inundated by this drainage effect, making China the quickest and worst affected region of the Earth. Concomitantly, other parts of the Earth at $\sim 30°N$ should be largely free of viral particles, unless it happens that such particles are incident as components within larger particles.

A direct demonstration that the general winter downdraft in the stratosphere occurs strongly over the latitude range 40° to 60° was given by Kalkstein.[14] A radioactive tracer, Rh-102, was introduced into the atmosphere at a height above 100 km and the incidence of the tracer was then measured year by year through airplane and balloon flights at altitudes ~ 20 km. The tracer took about a decade to clear itself through repeated downdrafts of the form shown in Fig. 12. Noting that the ordinate scale is logarithmic, the incidence of the Rh-102 is seen to be much greater in temperate latitudes than elsewhere, with the period January to March the dominant months.

The observed incidence of a radioactive tracer agrees closely with the well-known winter season of the viruses responsible for the majority of upper respiratory infections, including influenza. Figure 13, (taken from Communicable Disease Report CDR 83/49, Public Health Laboratory Service) shows the year-by-year incidence of respiratory syncytical virus, demonstrating a remarkable temporal concurrence with the radioactive data of Fig. 12. How we wonder is the almost clockwork regularity of RS infections to be explained otherwise? Unfortunately so little has been understood of the mode

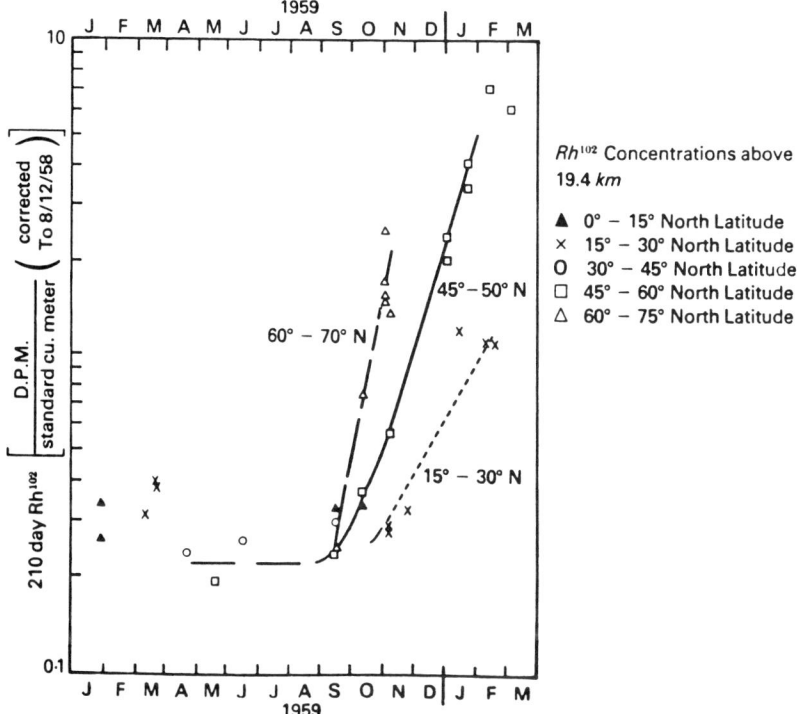

Figure 12.
The fall-out of Rh-102 at various latitude intervals from the HARDTACK atmospheric nuclear bomb which was exploded on 11th August 1958.

of attack of so-called infectious diseases that almost any form of hypothesis has come to be accepted in the past as an answer to questions of this sort. The truth is that, although the world may be extremely complex it is nevertheless extremely precise, with explanations every bit as clear-cut as that of the quantum mechanical analysis of the energy levels of the hydrogen atom being ultimately available for every phenomenon we observe.

Details of Vertical Incidence

It was remarked above that the world is extremely complex, and it is interesting to notice how true this can be, even for the first steps in the acquisition of small cometary particles by the Earth, for the Earth could cut just once on a unique occasion through a trail of evaporated particles, or it could cut periodically or irregularly at both closely and widely spaced time intervals. Besides which a pathogenic agent could be carried by a distribution of particles with varying sizes that descended through the terrestrial atmosphere in quite different intervals of time according to the discussion of the previous section. To recapitulate, particles with sizes ~ 10 μm fall under gravity everywhere over the Earth in only a few weeks. Particles with sizes ~ 1 μm fall in a few years, sooner in low latitudes than in high, while particles with sizes ~ 0.1 μm have a winter season from about December to March (in the northern hemisphere) when they are carried down through the lower stratosphere by mass movements of air, a process that occurs dominantly over the latitude range $\sim 40°$ to $\sim 60°$ and which would lead to a cyclically regular disease pattern like that of Fig. 13. To these previous considerations we must note that particle sizes can change not only due to water drop formation in the troposphere but to the aggregation of one particle with another produced by sticking effects should the particles become coated by acid as they pass through a sulphur layer of volcanic origin at heights of ~ 20 km. Thus time intervals could be appreciably affected by the varying exudations of SO_2 from volcanoes, as well as by the already mentioned effects of high mountain ranges. Finally, we have to consider the complexities that can arise at ground-level itself, complexities giving rise to the local variability in attack rates of a disease such as for Eton College in Fig. 7.

The epidemiologist observes the nett outcome of all these complexities, with the situation so scrambled together as to present an almost impossible problem of unscrambling, at any rate when the situation is treated empirically. Only with the aid of a model allied to observation can progress be made, as for instance the model of the winter downdraft in the stratosphere (Fig. 12) leading to an understanding of the pattern of RS infections (Fig. 13). It is a further consequence of this model that similar effects should occur in the southern hemisphere,

VIRUSES FROM SPACE AND RELATED MATTERS (1985)

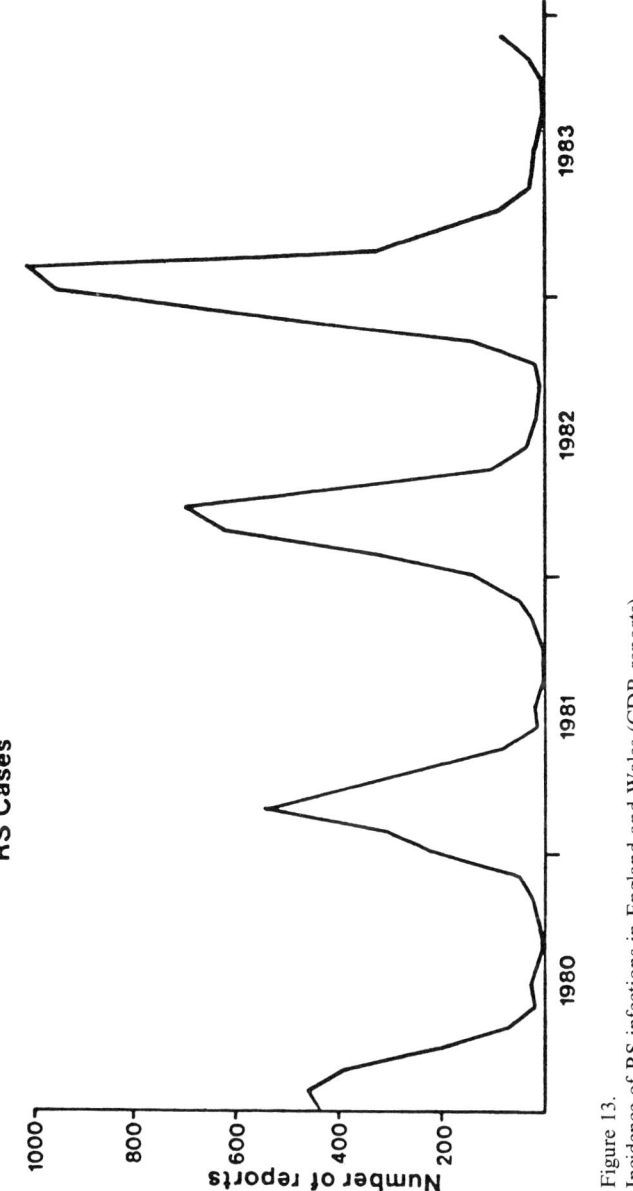

Figure 13.
Incidence of RS infections in England and Wales (CDR reports).

but six months different in phase because of the alternation of summer and winter in the two hemispheres. We do not have comparable data for RS infections, but influenza behaves similarly and data for influenza confirms the prediction of the model. Thus Hope-Simpson noted the phase variation in the occurrence of influenza across the continent of Africa over the period 1950–51, while we ourselves have assembled in Fig. 14 data issued by public health authorities in Sweden, Sri Lanka and Melbourne, Australia over periods ranging from 5 years to more than a decade. The global inference from the model is thus confirmed.

We have chosen to show the data for Sweden rather than Britain, not because there is any important difference between Sweden and Britain, but to bring out the point that the simple physical cold of winter is not a relevant factor. Sweden has a really cold winter, whereas Australia has a clement winter not much cooler than a Swedish summer. If simple exposure to cold were important, the effect would long ago have been demonstrated under controlled conditions in the laboratory, which it has not been.

Influenza A pandemics, following changes of virus type, do not fit the annual winter cycle in the manner of Fig. 14. Influenza pandemics fit readily into the model, however, with each major influenza type assigned to a separate accretion of virus from space, and with the viral particles being present in larger aggregates as well as individually by themselves. The larger particles fall through the atmosphere under gravity, those with sizes ~ 10 μm in a matter of weeks, those with sizes of ~ 1 μm in a year or two, and those existing by themselves with sizes ~ 0.1 μm taking a decade or more to reach ground-level. It is the latter extended period, characterised by so-called antigenic drift, possibly caused by solar ultraviolet light, that displays the effect of Fig. 14, after the earlier accretion of larger particles is over and done with.

Exactly where on the Earth the larger particles, responsible on this view for the initial outbreaks of an influenza pandemic, first reach ground-level is a matter of the vagaries of wind and weather, with the possibility that the larger particles reach

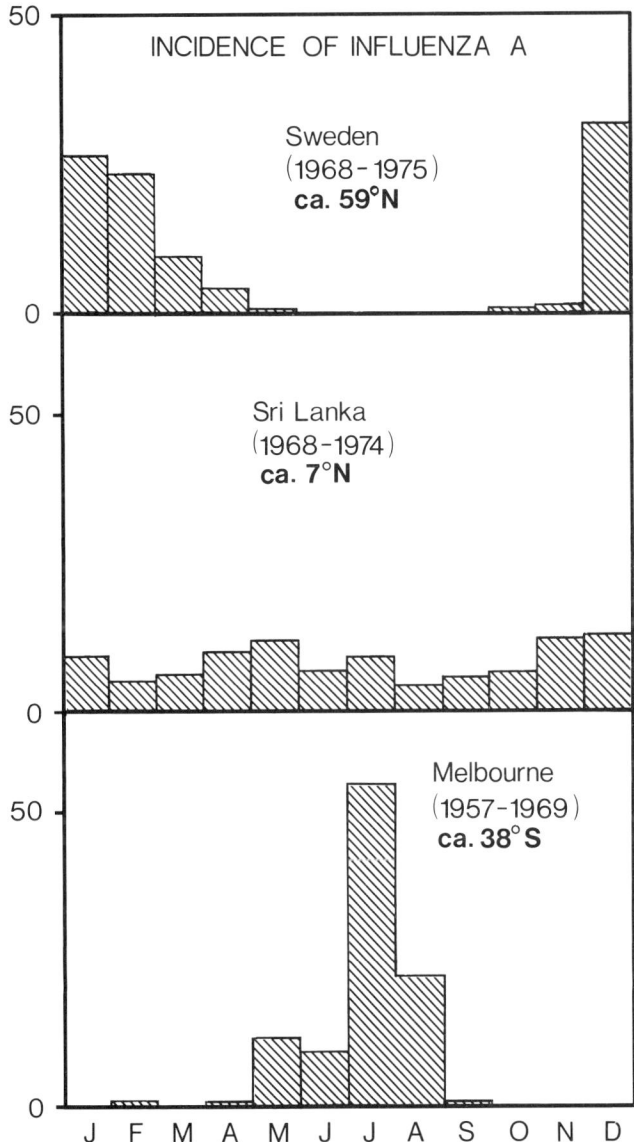

Figure 14.
Incidence of Influenza A in three separate countries.

ground-level more or less contemporaneously at widely-separated localities. If in accordance with the currently favoured person-to-person transmission hypothesis, one describes the first place of incidence as the 'focus' of the pandemic, the disease would appear to spread from the initial focus to other foci, perhaps separated geographically by large distances, with an amazing and quite inconceivable rapidity. Thus in the 1918/1919 pandemic the first outbreaks occurred within hours of each other in Boston, USA and Bombay, India, an impossibility for person-to-person transmission in days before air travel.[14a]

Also indicative of the short descent times of larger particles was the near contemporaneous outbreaks of the so-called Asian flu' pandemic of 1957-58, with similarities of timing that are again beyond reasonable possibility on the person-to-person transmission hypothesis, for example, the data in the following Table:

Table 1

Data on the 1957-58 'Asian' Influenza Pandemic

Place/Latitude		Onset	Peak	Decline	Ref.
U.K.:					
Tillingbourne, Kent	51°N	Sept. 2	1st wk Oct.	1st wk Nov.	(15)
St. Pauls, Kent	51°N	Early Sept.	Oct. 13	Early Nov.	(16)
Beckenham, Kent	51°N	Sept. 2	Oct. 14	Nov. 8	(17)
Ibstock, Leic.	52°N	Sept. 14	Oct. 12	Nov. 2	(16)
Cirencester, Glouc.	52°N	Sept.	Oct. 14	Nov. 4	(18)
Hawick, Scotland	55°N	-	Oct. 11	-	(19)
Collington, Cornwall	51°N	-	-	Nov. 6	(18)
U.S.A.:					
Cleveland, Ohio	40°N	Sept. 2	Oct. 13	Nov. 17	(20a)
Denver, Colorado	40°N	Early Sept.	1st wk Oct.	Nov. 17	(21)
U.S.S.R.:					
Moscow	56°N	-	Mid-Oct.	-	

The localities in the above Table are at nearly the same latitude. As expected, the effect of the lower height of the tropopause at higher latitudes caused a delay of some weeks, with the initial outbreaks in Alaska occurring at St. Pauls (57°N) on Oct. 11 and at Campbell Island (63°N) on Oct. 30.[20]

Consider next the situation at ground-level itself. It is a matter of experience that we do not normally snuffle raindrops up into the nose or gulp them into the mouth. So if the viruses causing upper respiratory infections fall down through the troposphere inside raindrops or snowflakes it may be wondered how we contract these diseases at all? Rain which impacts the face tends to drip off the end of the nose instead of entering the respiratory tract, possibly the reason for the possession of a nose. But rain does not end because all the water has fallen out of the atmosphere. Rain ends because falling droplets evaporate before reaching the ground. Droplets evaporating immediately in front of one's face, releasing viral particles into the air, would not be harmless because the released particles could then be breathed into the respiratory tract. It is therefore the end of a shower of rain that is dangerous, with the details highly local and irregular, thereby explaining why the incidence of breathable viruses at ground-level is often confined to small patches such as those responsible for the varying behaviours of school houses with respect to influenza, as in Fig. 7. The houses of Eton College which experienced unusually high attack rates were the ones where it happened that pupils were in exposed positions at just the moment when a shower of rain dried up, and conversely for the houses where attack rates were markedly low. It is also apparent why there can be no reproducability from one epidemic to another in the identities of the fortunate and unfortunate houses, the relation of particular places on the ground to the turbulent swirl of falling droplets being essentially a matter of chance on distance scales of the order of the precincts of a school. But not entirely a matter of chance on a scale of miles or on the scale which separates town and country. The efflux of heat from a large city could play a significant rôle in evaporating water droplets, and if the evaporation occurs high enough above the heads of city people they would be less at risk than country folk since country folk have no such self-protecting source of heat at their disposal. This may explain why attack rates of many diseases appear to be higher in the countryside than in neighbouring cities, as indicated in Fig. 3.

It must be rare for snowflakes to evaporate, because at the temperatures $< \sim 0°C$ at which snow is able to reach the ground without melting into rain, evaporation rates are low. Hence cold conditions with precipitation falling as snow should on the vertical incidence model be almost free from the danger of upper respiratory infections. Neither is heavy rain a dangerous condition. It is misty, drizzly weather that provides incoming viruses with the opportunity to become dispersed in the air close to ground-level. These expectations agree very well with popular lore, according to which damp is unhealthy but sharp cold weather is healthy. Shakespeare expressed the general lore when he wrote "... the winds ... have suck'd up from the sea, contagious fogs ...".

In the exceptionally cold weather of January and February 1985 it was widely noticed that Influenza A was essentially absent throughout the world, and that in the U.K. there was an atypical absence of upper respiratory infections generally. We ourselves predicted that once damper, warmer weather set in, as it did in late February, there would be a sharp rise of such infections, as indeed actually happened, with Influenza A at last appearing over the whole latitude belt ranging from the U.S.S.R. to the western states of America. In effect, the exceptionally cold weather held off the infections which normally occur in January and February.

The above discussion has been biased to suit the situation in northern temperate latitudes. If one lives in desert conditions, other factors would be seen to be important. Precipitation in deserts tends to be one thing or the other, either heavy or absent, with little of the damp, drizzly weather of northern latitudes. Viruses falling from the atmosphere would mostly reach ground-level therefore without constituting a serious *immediate* threat to a desert population. Once on the ground they would largely stay there, however, instead of being washed away in streams and rivers, to be stirred-up into the atmosphere again by winds. Windy periods with sandstorms would thus be the times when upper respiratory infections appear, an expectation which according to our somewhat fragmentary knowledge of desert conditions seems to be correct.[23]

The circumstance that in a vertical incidence model viruses could come upwards from the ground as well as downwards from the atmosphere raises still further complications. While most of the falling viruses to reach ground-level in regions other than deserts will be washed away by streams and rivers, some viral particles remain suspended in the subsurface water table, and some would accumulate in fresh-water lakes. Evaporation provides a channel whereby water-embedded viruses could subsequently be released into the atmosphere near ground-level. There are many subchannels whereby evaporation occurs, indoors from the domestic water supply at all times of the year, and outside mostly in spring and summer. A high volume subchannel lies in evaporation from the leaves of trees. This particular subchannel is especially interesting because of the tendancy of people to plant trees around their houses, and because the evaporation process cools the air around trees due to the absorption of heat necessary to supply the latent energy of evaporation. The cooled air, being denser in summer than warm surrounding air, falls immediately to ground-level thereby dumping any viruses it may contain on to unsuspecting folk living or picnicking in the shelter of trees.

The biological system of a tree probably filters out most viruses, particularly those with an affinity for cellulose, but a minority may well be expelled both by trees and foliage of all kinds into the atmosphere. The minority that are not filtered then appear as typical summer diseases. Figure 15, taken from Communicable Disease Report 83/24, shows reported cases of parainfluenza from 1980 to 1983. Year after year, cases begin a steep rise in April, attain a maximum in June, and fall essentially to zero in September, in correspondence with the annual growing cycle of plants. It is hard to imagine any other process that could give a cycle so repeated and so perfectly timed.

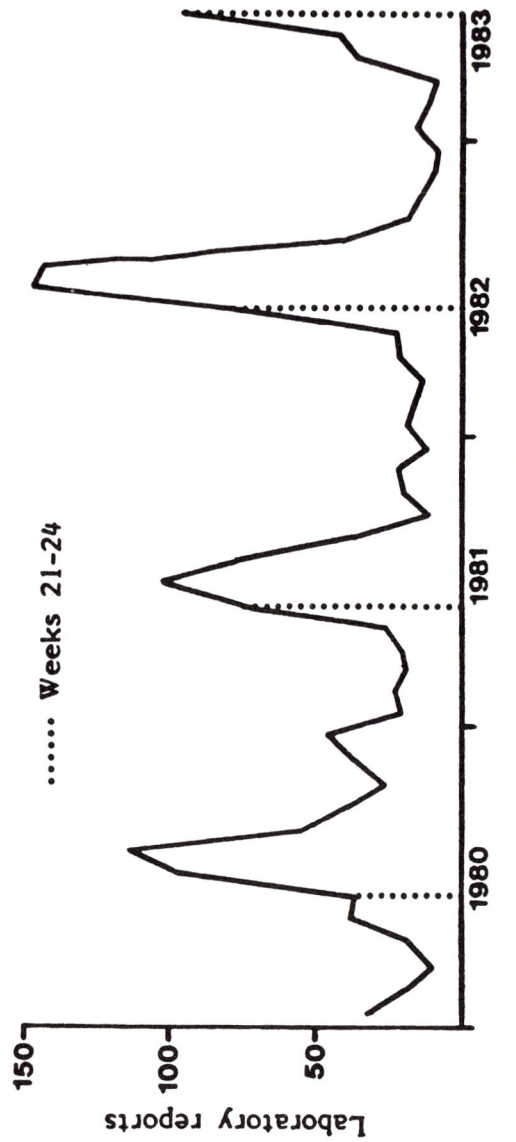

Figure 15.
Incidence of Parainfluenza type 3 (CDR reports).

Conclusions

On the vertical incidence theory localities of exceptionally high incidence are to be expected at ground-level. Ironically, such regions appear at first sight to give credence to the person-to-person transmission theory, because people living in a locality of high vertical incidence have an impression that they are infecting each other. Likewise whenever any of us contracts an upper respiratory infection after being in contact with an early victim of a similar infection we tend to believe we have 'caught' the infection from the earlier victim. Such perceptions lack quantitative support, however, for whenever they are examined with a little care, discrepancies for the person-to-person theory soon emerge.

Data collected on all scales, ranging from the individual surgeries of general practitioners to worldwide patterns of disease show overwhelmingly that most acute upper respiratory infections arise from vertical incidence. The data could be extended greatly at comparatively little effort and cost, as for instance the determination of the attack rates of particular diseases as a function of population density. The reason why such data is not precisely analysed is we think psychological, because it would instantly destroy conformist opinion, which in all branches of science avoids as far as possible confrontations with awkward facts. Yet the benefits to public health from a clear understanding of the cause of acute upper respiratory tract infections could be very great. With the modes of incidence known for various diseases only quite simple precautionary measures could well serve to reduce morbidities very appreciably, with consequent benefits economically for the community at large as well as for individuals personally.

References

[1] R.C.G.P.P.H.L.S. (1981) Influenza in families: Preliminary report based on the winter of 1973-4, *Journal of the Royal College of General Practitioners*, **27**, 19-26.
[2] M.R.C. A collaborative study of the aetiology of acute upper respiratory tract infections. 1961-4. *British Medical Journal* (1965) **2**, 319-26.
[3] Pfeiffer, R. (1892) *Dtsch.Med.Wschr.*, **18**, 28.

4. Hoyle, F. and Wickramasinghe, N.C., (1979) *Diseases from Space*, (J.M. Dent, London).
5. Hope-Simpson, R.E. and Sutherland, I., (1954), Does Influenza spread through the household, *Lancet*, **1**, 721.
6. Hope-Simpson, R.E., (1979) Epidemic mechanisms of Type A Influenza, *Journal of Hygiene*, Cambridge, **83**, 11-26.
7. Philip, R.E., et al., (1961) Epidemiological studies on influenza in familial and general population groups, 1951-6, *Am.J.Hyg.*,**73**, 123-137.
8. Mann, P.G., et al., (1981) A five-year study of Influenza in families, *Journal of Hygiene*, Cambridge, **87**, 191-200.
9. Dr. J. Skehel (1985) World INFLU Centre, London (Personal communication).
10. Sekanina, Z., (1970), *Icarus*, **13**, 475.
11. Arrhenius, S., (1908), *Worlds in the Making*, (Harper & Row, New York and London).
12. Hoover, R.B., Hoyle, F., Wickramasinghe, N.C., Hoover, M.J. and Al-Mufti, S., (1985) *Cadiff Astrophysics and Relativity Preprint* 114; in *Earth, Moon and Planets*, in press.
13. Hope-Simpson, R.E., (1981) The influence of season on type A Influenza, *Journal of Hygiene*, Cambridge, **86**, 35.
14. Kalkstein, M.I., (1962) *Science*, **137**, 645.
14a. Weinstein, L., (1976) Influenza-1918 a revisit?, *New England J.Med.*, **294**, 1058-1060.
15. Watson, G.I., (1960)*Journal of College of General Practitioners*, **3**, 44-79.
16. Woodall, J., Rowson, K.E.K. and Macdonald, J.C., (1958) Age and Asian Flu, 1957, *British Medical Journal*, **2**, 1316-18.
17. Fry, J., (1958) Influenza A Asian 1957, *British Medical Journal*, **1**, 259.
18. Respiratory Diseases, (1958) in *British Medical Journal*, **1**, 110-114.
19. Mcgregor, R.M., (1957) *British Medical Journal*, **2**, 1058-1059.
20. Philip, R.E., et al., (1959) Observations on Asian Influenza on two Alaskan Islands, *Public Health Report*, **74**, 737-745.
20a. Jordan, W.S., et al., (1958) A Study of Illness in a Group of Cleaveland Families, *Am.J.Hyg.*, **68**, 190-212.
21. Meiklejohn, G., (1983) *The J. of Inf. Diseases*, **148**, 775.
22. Dunn, F.L., (1958) Asian Influenza Pandemic, *J.A.M.A.*, **166**, 1140-1148.
23. Hassan, Prof. M.H.A., (1983) Private communication.

2

THE ANATOMY OF A CONFERENCE EPIDEMIC (1983)

There are numerous accounts in the literature of explosive outbreaks of acute upper respiratory disease in so-called closed communities – e.g. in airplanes, ships, boarding schools, army barracks and hospitals. In a typical story an 'index case', essentially the first case to be noticed, is identified and it is argued that this individual, who supposedly introduces the infective pathogen from outside, then transmits the disease sequentially to susceptible members in close contact within the closed group. We ourselves came to have serious doubts about the validity of such explanations after our own efforts in 1979 to study the behaviour of an influenza epidemic as it affected a large number of children in independent schools (including boarding schools) in England and Wales.[1,2] In our investigation, absence from school was taken as a measure of the attack rate on any particular day. We accumulated a substantial body of evidence that militated against the hypothesis of person-to-person transmission of influenza. All the data as were available to us were readily explained if the viral pathogen (or an activating component of it) settled on the Earth's surface from above, the fine details of the observed attack patterns being determined essentially by meteorological factors that control incidence. Consistent with this picture we found that some schools experienced exceptionally high attack rates, whereas others, sometimes in close proximity, were virtually unscathed. Had we chosen individual schools with high attack rates in isolation from the rest of the data we might easily have convinced ourselves that the standard explanation must be true.

At a recent international seminar held in Sri Lanka, in which we were both involved, there occurred an epidemic of acute upper respiratory disease (diagnosed as such by attendant physicians) that we were able to observe. The manner in which the disease propagated explosively seemed at first sight to provide clear evidence for person-to-person transmission of some causative pathogen. A closer scrutiny revealed a different story that we now describe.

The time was December 1982; the place Colombo, Sri Lanka. Some thirty-three international scientists and a roughly equal number of Sri Lankan scientists had assembled to take part in a seminar on the theme 'Fundamental Studies' organised by the Institute of Fundamental Studies, Sri Lanka. The participants (some of whom were accompanied by spouses and families) resided at the Lanka Oberoi Hotel Colombo, a modern air-conditioned hotel, for the entire duration of the seminar. The lectures and formal discussions took place in rather lofty air-conditioned lecture rooms at the Bandaranaike Memorial International Conference Hall (BMICH), situated about 3 miles away from the hotel. Although the conference proper was from Thursday 2 December–Saturday 11 December, many participants arrived in the hotel a few days earlier and stayed on for some days after the 11th.

The so-called index case in our epidemic was seen to be an astronomer who arrived from New York on 30th November. He arrived in good health, but on Friday 4th December went down with an acute attack of upper respiratory illness. In some ways the illness was not unlike a severe attack of a common cold, but there were some notable differences. The onset of the attack was accompanied by a slight fever; there was less nasal discharge and irritation than normally characterises a cold; and there was also an acute soreness of the throat somewhat untypical of a common cold. Three to four days later (on Tuesday December 7th and Wednesday 8th) the present authors become stricken with the same affliction, and shortly afterwards we learnt that many of the other participants also went down. The illness ran a benign course with an acute phase lasting some 3–4 days, followed by a more protracted and less troublesome phase of somewhat variable duration. Those participants who chose to consult a medical practitioner were prescribed erythromycin, 250 mg four times daily. It was not clear, however, whether or not the administration of this antibiotic had any effect at all on the course of the disease. Significantly perhaps the disease was not seen to pass from infected participants to their spouses and families including an infant in one instance. At a rough estimate about 35 of the individuals connected with the seminar seemed to have fallen victim by the time the formal sessions ended at about 13:00 hours on Saturday 11th December. Incidentally, we noticed that at the concluding session a noticeably high proportion of the audience (comprised of nearly 1000 people) seemed to be coughing during the lectures.

On Sunday 12th December the conference participants visited the University of Peradeniya near Kandy, a city nestled in the hills of central Sri Lanka some 70 miles or so from Colombo. Talking to the Medical Officer of the campus we learnt that an influenza-like illness of the type we had experienced was generally quite widespread in Colombo during the preceding week, but that it did not seem to have reached Kandy. According to the Medical Officer, no cases of acute upper respiratory disease had come to his notice amongst the campus population in Peradeniya. This conversation led us to seek further details and information regarding the epidemic that we had experienced in Colombo.

Back in Colombo we first obtained school absence data for 6th grade students (average age 12 years) in several schools in the vicinity of both the Lanka Oberoi Hotel and the Conference Hall. Fig.2.1 shows this record for the period 15 November 1982–15 December 1982 for a total student population of 1257 in six schools. Note that a dramatic rise in absences occurred between 1 December and 3 December, with an epidemic peak occurring on 7 December, close to the date when the conference epidemic was also at its peak. The attack rate (implied by absences) varied significantly between the several schools but there was only a relatively small dispersion of the peak date. (This data is shown in Table 2.1.) In each of the schools the difference between absence patterns of the 6th and 7th grade students (the only two age groups we looked at) was almost negligible. These results correspond well with our earlier findings in the U.K. influenza epidemic of 1978/79.[1] Of

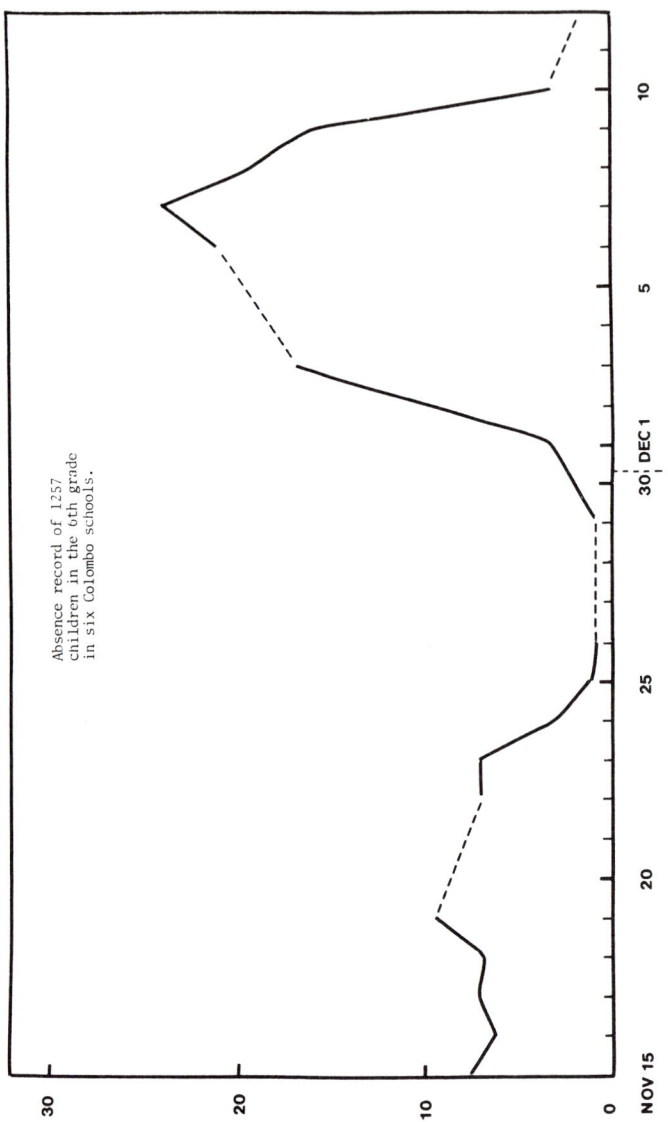

Figure 2.1.
The absence record of 1257 children in the 6th grade in six Colombo schools (November and December 1982).

course we do not possess unequivocal evidence that the same affliction we ourselves experienced was the cause of the majority of absences in Colombo schools during November and December, but enquiries made from a cross-section of teachers and pupils strongly suggested that this was so.

Table 2.1

Tabulation of the maximum percentage absence and the date of peak absence for several schools

	School	Maximum % Absent	Date of Peak Absence
1	DSS	41.3	Dec 8
2	VKV	9.3	Dec 7
3	DBV	9.7	Dec 7
4	NV	25.8	Dec 8
5	CIS	20.0	Dec 6
6	BS	46.3	Dec 3

Another simple method of investigation was to examine the record of antibiotic prescriptions dispensed at a well-known Colombo chemist 'Osusala'. The Chief Pharmacist at 'Osusala' kindly supplied us with numbers of dispensations of antibiotics (mainly erythromycin) which are normally given for acute upper respiratory symptoms on a day to day basis. This chemist is open 24 hours a day all week including Sundays and Public Holidays. A first glance at his data showed that conspicuously high numbers of dispensations were recorded for the week November 28–December 4 compared with neighbouring weeks before and after this period. This weekly average number dispensed for October, November and December was found to be $\bar{x} = 319.86$, excluding the exceptional week November 28–December 4. The sample standard deviation about this mean was calculated to be $\sigma = 27.76$. We next proceeded to calculate the departure Δ from the mean weekly dispensation rate, week by week, from Sunday October 31, 1982 to Saturday December 25, 1982. Fig. 2.2 shows a histogram in which this departure Δ is plotted relative to the standard deviation σ. We note that the week November 28 – December 4 is characterised by an increase of the

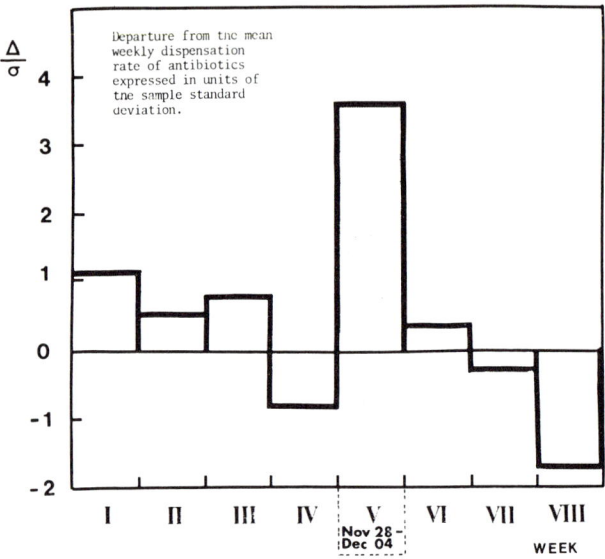

Figure 2.2.
The departure from the mean weekly dispensation rate of antibiotics for upper respiratory and ENT infections, expressed in units of the sample standard deviation.

antibiotic dispensation above the mean by 3.61 standard deviations. (The week December 19–December 25 has a lower-than-average antibiotic dispensation number by 1.72 standard deviations, an effect that might perhaps be attributed to a Christmas holiday spirit conferring a higher degree of psychological tolerance to minor afflictions!). Clearly, the November 28–December 4 increase of Δ/σ is highly significant in confirming our supposition that the conference epidemic was part of a much wider phenomenon.

We present these observations as a cautionary tale directed to those who are apt to seek an easy proof of person-to-person transmission from apparently dramatic manifestations of epidemics in small groups. We ourselves might easily have reached a similar conclusion both on the present occasion as well as earlier in 1978/79 had we confined our attention only to those apparently 'closed communities' that reported exceptionally high attack rates sharply localised in time. The fact remains, however, that few such communities can be regarded as closed to any reasonable degree. It is equally a fact that we are nearly always forced to breath the outside air and all that goes or comes with it. In the case of our recent conference epidemic, there clearly was an upper respiratory tract pathogen that was blowing through the Colombo area exactly at the time of the December conference. When the present authors drove out of Colombo to visit the Royal Botanic Gardens at Peradeniya on the morning of 1 December we had already noticed that the outskirts of Colombo were shrouded in a low-lying mist. In our view it is precisely this type of meteorological condition that favours the spread of a space-incident respiratory pathogen. Our American colleague, the best candidate to be considered as an 'index case' (in the standard theory) happened perchance to be the first susceptible person to inhale this pathogen in the course of early morning walks around Colombo between November 30 and December 3.* The other victims (including ourselves) were also undoubtedly infected in a similar manner along with many thousands of school children and others in the Colombo area (see Figs. 2.1 and 2.2). This somewhat unusual case history might serve to illustrate the importance of connecting the patterns of an epidemic discovered within a 'closed community' with relevant information from the wider world outside.

We thank Mr. H. P. Saranadasa and Mr. J. A. Mariampillai (Pharmacist at Osusala) for assistance in collecting and supplying relevant data.

---

* This individual was accustomed to take such early morning walks because of the timeshift between New York and Colombo.

References

[1] Hoyle, F. and Wickramasinghe, C. 1979, *Diseases from Space* (J. M. Dent and Sons).

[2] Hoyle, F. and Wickramasinghe, C. 1980, *Space Travellers: the Bringers of Life* (Univ. Coll. Cardiff Press).

3

THE PLAGUE AT ATHENS – WAS IT SMALLPOX? (1981)

Introduction

The plague of Athens broke out in the summer of 430 BC. It is said by Livy to have lasted for four years. The eye-witness description given by Thucydides in *The Peloponnesian Wars* presents a riddle of medical diagnostics that has for long remained unsolved. Indeed there has not been a great deal of progress in the matter since H. A. J. Munro wrote in 1860:

> "Physicians–English, French, German–after examining the symptoms, have decided that it was each of the following: typhus, scarlet, putrid, yellow, camp, hospital, jail, fever; scarlatina magna; the Black Death; erysipelis; smallpox, the oriental plague; some wholly extinct form of disease. Each succeeding writer at least throws doubt on his predecessor's diagnosis."

Our own interest in the Plague of Athens arose from our belief that the primary causative agents of infectious diseases arrive at the Earth from space[1]. It would be to our advantage in this respect if the last of the possibilities mentioned by Munro were correct, for it is a consequence of our point of view that infectious diseases should be considerably variable on a long time-scale of two thousand years. We would therefore prefer the Plague of Athens to have been a disease unknown to modern medical science. Precisely because our motivation is strongly in this direction, we have 'played the devil's advocate', by giving modern diseases the benefit of the doubt in ambiguous or uncertain comparisons with Thucydides. Yet even so we have satisfied ourselves that the fit is unconvincing for each disease which has been suggested, except possibly for smallpox. In this article we attempt to show why smallpox is a possibility that cannot easily be dismissed.

English translation by Benjamin Jowett of paragraphs 49 and 50 from Book II of Thucydides' The Peloponnesian Wars.

"The season was universally admitted to have been remarkably free from other sicknesses; and if anybody was already ill of any other disease, it finally turned into this. The other victims who were in perfect health, all in a moment and without any exciting cause, were seized first with violent heats in the head and with redness and burning of the eyes. Internally, the throat and the tongue at once became blood-red, and the breath abnormal and fetid. Sneezing and hoarseness followed; in a short time the disorder, accompanied by a violent cough, reached the chest. And whenever it settled in the heart (stomache), it upset it; and there were all the vomits of bile to which physicians have ever given names, and they were accompanied by great distress. An ineffectual retching, producing violent convulsions, attacked most of the sufferers; some, as soon as the previous symptoms had abated, others, not until long afterwards. The body externally was not so very hot to the touch, not yellowish but flushed and livid and breaking out in blisters and ulcers. But the internal fever was intense; the sufferers could not bear to have on them even the lightest linen garment; they insisted on being naked, and there was nothing which they longed for more eagerly than to throw themselves into cold water; many of those who had no one to look after them actually plunged into the cisterns. They were tormented by unceasing thirst, which was not in the least assuaged whether they drank much or little. They could find no way of resting, and sleeplessness attacked them throughout. While the disease was at its height, the body, instead of wasting away, held out amid these sufferings unexpectedly. Thus, most died in the seventh or ninth day of internal fever, though their strength was not exhausted; or, if they survived, then the disease descended into the bowels and there produced violent lesions; at the same time diarrhea set in which was uniformly fluid, and at a later stage caused exhaustion, and this finally carried them off with few exceptions. For the disorder which had originally settled in the head passed gradually through the whole body and, if a person got over the worst, would often seize the extremities and leave its mark, attacking the privy parts, fingers and toes; and many escaped with the loss of these, some with the loss of their eyes. Some again had no sooner recovered than they were seized with a total loss of memory and knew neither themselves nor their friends.

The character of the malady no words can describe, and the fury with which it fastened upon each sufferer was too much for human nature to endure. There was one circumstance in particular which distinguished it from ordinary diseases. Although so many bodies were lying unburied, the birds and animals which feed on human flesh either never came near them or died if they touched them. This is the evidence: there was a manifest disappearance of birds of prey, which were not to be seen either near the bodies or anywhere else; while in the case of the dogs, what happened was even more obvious because they live with man."

Concordances between Thucydides and a modern description of confluent smallpox (for example, Osler's Principles and Practice of Medicine*)

Thucydides: ...victims in perfect health (were) all in a moment seized first with violent heat in the head and with redness and burning in the eyes.

Osler: The temperature rises quickly and on the first day may be 103° or 104°. Intense frontal headache (is) a very constant feature...the face (is) flushed and the eyes bright and clear.

Thucydides: Internally, the throat and tongue at once became blood-red, and the breath abnormal and fetid.

Osler: The eruption may be present in the mouth, usually the pharynx and larynx are involved. The odour of a smallpox patient is very distinctive even in the early stages...

(There is a discrepancy of timing here, however, with the 'at once' of Thucydides implying the pre-eruptive invasion phase.)

Thucydides: ...hoarseness followed.

Osler: ...the voice is husky.

Thucydides: ...there were all the vomits of bile...ineffectual retching...

Osler: ...vomiting (is) a very constant feature.

---

* The quotations are from the 1944 edition of Osler, when smallpox was still fairly common.

Thucydides:	The body externally was not so very hot to the touch, not yellowish, but flushed and livid...
Osler:	The scarlatina rash may come out as early as the second day...often purpuric...
Thucydides:	...the sufferers could not bear on them the lightest linen garment.
Osler:	Itching is present, and this is a troublesome symptom...
Thucydides:	They were tormented by unceasing thirst...
Osler:	Usually there is much thirst.
Thucydides:	While the disease was at its height, the body, instead of wasting away, held out amid these sufferings unexpectedly.
Osler:	With the appearance of the eruption the symptoms subside... Occasionally the temperature falls to normal, and the patient may be comfortable.
Thucydides:	...the disease descended into the bowels ...diarrhea set in...and at a later stage caused exhaustion, and this finally carried them off with few exceptions.
Osler:	...the pulse gets feebler...there is sometimes diarrhea, and with these symptoms the patient dies.
Thucydides:	...leave(s) its mark, attacking the privy parts.
Osler:	Inflammation of the testes and ovaries may occur.

Thucydides: Some escaped with the loss of (the use of ?) fingers and toes, some with the loss of (the use of ?) their eyes.

Osler: Local gangrene in various parts may occur. Arthritis is met...purulent conjunctivitis is common in severe cases. Unless great care is taken a diffuse keratitis occurs which may go on to ulceration and perforation.

Thucydides: Some again had no sooner recovered than they were seized with a total loss of memory and knew neither themselves nor their friends.

Osler: Post-febrile insanity is met...

Issues on which the Precise Meaning of the Original Greek is Critical

It is important to know whether the eruption occurred in the skin, or on the skin. Benjamin Jowett gives the skin "breaking out in blisters and ulcers", while D. L. Page, in a scholarly discussion[2] of the medical terms used by Thucydides, has the skin "efflorescing" into blisters and lesions. Because of the relevance of this issue to whether the disease could have been measles, as favoured[3] by J. F. D. Shrewsbury and by Page, we asked Dr. Humphrey Palmer of University College, Cardiff, for a literal translation of the original Greek. He kindly provided us with:

> "...the skin was rather reddish and livid and flowered out into pustules and sores."

Dr. Palmer's translation supports smallpox, not measles.

According to both Jowett and Page, victims died mostly on the seventh or the ninth day, whereas deaths from smallpox occur mostly towards the end of the second week. However, what Thucydides actually wrote on this matter of timing was very peculiar. Dr. Palmer kindly supplied us with the following literal translation of the relevant Greek sentence:

> "While the disease was at its height the body withstood prostration beyond what would be expected, so that they died in most cases as *niners and seveners*."

Translators have uniformly interpreted the conjunction 'and' as equivalent to 'or', and so have arrived at the conclusion that victims died either in seven days or in nine days. This leaves as a mystery what happened on the eighth day. If there was a remission on the eighth day, it is strange that Thucydides did not give more emphasis to such a remarkable feature. One 19th century translator, faced with this logical problem, even changed nine to eight (!), and all translators have inverted the ordering of nine and seven, even though Thucydides has been described as a writer of "minute and scientific accuracy". Presumably he put nine before seven for some good reason. The question is what?

The more usual meaning of the conjunction 'and' is 'plus', so that in this predicament it is reasonable to ask if there is a sense in which niners *plus* seveners can be given a meaning. Dr. Palmer informs us that in the fifth century B.C. there had been a debate as to whether weeks should be counted in groups of seven days or of nine days. Even a century or more after Thucydides, market days were held in Rome in weeks of nine days.

Suppose that what Thucydides wished to say was that 'most cases died in about two weeks', in agreement with smallpox. Being a writer of "minute accuracy", he notices that the reader might worry about which kind of a week he intended. So to remove ambiguity he specifies a week of each kind–this can be done without violence to the facts because the interval is in any case only an approximation, *about* two weeks. Finally, to prevent 'and' being misinterpreted as 'or', he inverts the ordering–a niner and a sevener–which usage we speculate would have been known to contemporary readers. Far-fetched? Possibly, but the alternative appears much worse!

Be this as it may, the usual translation of 'seven or nine days' is heavily interpretive, and so cannot convincingly be used as an objection to smallpox.

Symptoms of Smallpox not Mentioned by Thucydides

All modern descriptions of smallpox emphasise that patients complain during the invasion of lumbar pains, which Thucydides does not mention. Indeed, Thucydides emphasised that symptoms moved progressively downward through the body: "For the disorder which had originally settled in the head passed gradually through the body..."

Thucydides refers to the skin only in a few sentences. While these sentences agree well enough with smallpox–"the skin flowered out into pustules..."–there is an imbalance between the restricted emphasis which Thucydides gives and the greater emphasis given by modern writers. This difference can perhaps be plausibly explained by the circumstance that Thucydides was himself a sufferer from the disease, whereas modern descriptions are not subjective. To an external observer, it is the skin that is visible and which naturally claims the attention, whereas the sufferer is afflicted by the internal depredations of the disease–Thucydides repeatedly emphasises the *internal* fever.

What cannot be explained in this way is that Thucydides does not mention disfigurement as an after effect of the disease. This omission has been considered by some commentators to rule out smallpox decisively.

Assessment of the Smallpox Hypothesis

We have not found any positive statement of Thucydides to be in serious conflict with the smallpox hypothesis. Although several of the concordances are fairly trivial, at least two of them are striking, namely "the body holding out unexpectedly" in the mid-course of the disease, and the description of the last feeble stages–"this finally carried them off with few exceptions."

In a recent paper, A. J. Holladay and J. C. F. Poole conclude otherwise[4]. They argue that in paragraph 50 Thucydides says that the disease attacked birds and dogs, whereas smallpox does not do so. Unlike paragraph 49, which is replete with positive statements, paragraph 50 is guardedly inferential, however. Nowhere does Thucydides actually say that he saw other animals attacked by the same disease as man. His sentences were carefully hedged: "This is the evidence: there was a manifest disappearance of birds of prey."

From our point of view, different causative agents from space could have caused different diseases contemporaneously in animals and in man.

The case against smallpox rests on omissions from Thucydides, of which the omission of lumbar pains from the early phase is the clearest example. The problem is to know how heavily to weight this negative evidence. Modern descriptions represent the averaged observations of many doctors who were not sufferers from the disease. Thucydides was one man, not a doctor as he emphasises, and he was himself a victim:

> "I shall speak the symptoms (of the disease)... I shall describe them clearly as one who was attacked myself, and witnessed the sufferings of others."

Smallpox contains many variations. Thus from Osler:

> "Smallpox also has its peculiar kinds, which take one form during one series of years, and another during another."

The obvious danger in making a diagnosis on negative evidence is that Thucydides weighted his description from his own experience of the disease. Perhaps in his case the lumbar pains were not so noticeable, or perhaps his memory of them became obliterated by later, more intense suffering.

None of this explains the omission of disfigurement, however, which remains the single strongest objection to the Plague of Athens being smallpox. Yet how strong is it? Thucydides may have been disfigured himself. How if an understandable sensitivity caused him to omit such a highly personal aspect of the disease?

Classical scholars with whom we have discussed this possibility have unanimously rejected it. Thucydides' statement: "I shall speak the symptoms–I shall describe them clearly..." is for them a guarantee that he told not only the truth but the whole truth. But if so, how are we to interpret the opening sentence of paragraph 50.

> "The character of the malady no words can describe and the fury with which it fastened on each sufferer ..."

Unless we interpret this sentence as mere rhetoric, Thucydides tells us that there were aspects to the horror which went beyond paragraph 49. He implies that the omissions are to be attributed to his inadequancy as a writer, but could this really have been so?

Unlike our friends, who were somewhat shocked at the suggestion that Thucydides might have deliberately concealed something, we cannot think it much of an intellectual sin for a man, keenly aware of a cruel disfigurement, to spare himself the indignity of describing it. Later in *The Pelopponesian Wars* Thucydides describes his own disgrace at Athens with great objectivity, but we doubt the issue of his disgrace could compare psychologically with a severe disfigurement.

There is a final point that could be raised as an objection to smallpox. We know that the Plague of Athens did not last for more than four years, and there is no evidence for smallpox existing anywhere else in Europe from the 5th Century B.C. to the dawn of the Christian era. Our more recent experience of smallpox is that the virus, no matter where it comes from, tends to form a stable reservoir in the human population that persists for several centuries. Indeed, the smallpox virus is so robust that it can probably survive outside a human quite easily for decades. Why then was the presumptive virus that caused the Plague so short-lived in its persistence? Is the degree of robustness of the smallpox virus a variable factor, so that we could postulate a more fragile virus in 430 B.C? Or, could it have been that the population density of susceptibles remained below the critical value for establishing a long-lived reservoir? We simply do not know.

Where then does this leave us? Pretty much where we began. The case for smallpox is by no means compelling, but not bad either.

References

[1] Hoyle, F. and Wickramasinghe, N. C., *Diseases from Space* (J. M. Dent and Sons, 1979).
[2] Page, D. L., 'Thucydides' Description of the Great Plague of Athens', *Classical Quarterly*, N. S. iii (1953).
[3] Shrewsbury, J. F. D., *Bull. Hist. Med. xxiv* (1950).
[4] Holladay, A. J. and Poole, J. C. F., 'Thucydides and the Plague of Athens', *Classical Quarterly*, **29,** 282 (1979).

4

INFLUENZA – A GENETIC VIRUS WITH AN EXTERNAL TRIGGER (1980)

Introduction

In the winter of 1977–78 we carried out an investigation of influenza epidemics which had occurred in certain Welsh and English schools[1]. About 25,000 pupils were involved, a large enough number to ensure adequate uniformity of the diagnostic procedures of school matrons and medical officers. The data showed that the incidence of cases depended scarcely at all on human contact but on the locations where the victims happened to be positioned at particular moments. The data indicated that the main causative agent of influenza is airborne, and that it settles from the atmosphere in a highly patchy distribution, analogous in its vagaries to smoke rings swirling around in a room.

This conclusion was consistent with our view that the whole biochemical apparatus of life is of external cosmic origin . We attributed the sudden onset of influenza pandemics, sometimes contemporaneous at widely different places on the Earth, to the arrival from space of a new sub-type of the virus.

A Fully-fledged Virus or a Trigger?

Several friends, particularly Prof. John Watkins of the University Hospital of Wales, Cardiff, and Dr. J. B. Selkon of Newcastle General Hospital, suggested that our views would be more attractive (less unpalatable!) if the need for the arrival from space of the complete influenza virus was replaced by the arrival of a DNA or RNA fragment, a viroid, which acted in some ways as a trigger for the disease. Such an idea has been published by Irvine *et al*[2] who otherwise took over our theory without substantive change.

The need to make the theory more plausible did not, however, seem to us as necessary as it appears to have been to critics who argued that viruses are uniquely related to their hosts, and so are most unlikely to have arrived from space. Although this criticism has been repeated many times it never fails to surprise us, for the reason that it is obviously quite untrue. So far from human influenza virus being uniquely specialised to human cells, it is a well-known fact that the virus can be cultured in calf and canine kidney cells for example, and even in the cells of an animal of an entirely different taxonomic class–chick embryo.

It is true that the lipid coat of a virus is very similar to the lipid membrane of the host cell, but this too is a trivium, since the virus derives its coat from the host. Thus the lipid coat of the human influenza virus takes on the characteristics of the lipid membranes of calf or canine kidney cells when cultured in them[3]. The same property applies for other viruses[4]. A man does not acquire the identity of another simply by robbing him of his jacket.

A better argument for replacing virus by viroid is that viroids are highly resistant to destruction[5]. The fragility of the influenza virus has always been a problem, which we have sought to cope with by supposing the virus to be coated in a protective matrix. The efficacy of a protective matrix has some experimental support[6], so that even this better argument did not seem decisive.

In the past we preferred the incidence of fully-fledged virus, because the viroid idea seemed to raise more questions than it answered.

Where then does the virus itself come from? The answer to this question would have to be that the virus lies already latent in the host cell, unable to multiply until the viroid arrives to supply a trigger for replication to begin. But then one has the mathematical difficulty that, unless a considerable fraction of respiratory cells contain lurking virus (in which case one might think the disease was there already) the probability of a viroid chancing to enter one of the cells where the virus just happened to be present would be small. Many more incident viroid particles would therefore be needed (than for the arrival of fully-fledged virus) to give a reasonably high probability of contracting the disease.

There are other problems. Suppose a latent virus particle to be triggered into operation inside a particular cell. When the resulting new viral particles leave the cell to invade other cells, do they carry the trigger with them, or are the new viral particles mysteriously no longer latent? Furthermore, what in a viroid theory causes sudden world-wide jumps in the antigenic properties of the latent virus? To answer questions such as these more and more *ad hoc* hypotheses seemed necessary, making us think the viroid idea unattractive.

The purpose of this article is to suggest a radical step which overcomes these former misgivings.

Genes in Pieces

Our present suggestion has its origin in the crucial discovery that the genes of eucaryotic cells are each assembled from a number of bits of the genome[7,8]. This we think precludes the use of standard initiator and terminator codons, as occurs in procaryotic cells. To keep order in the arrangement of bits of the genome it would seem necessary for the bits to be 'addressable' individually. The difference between a cheap hand calculator and the more expensive hand computers will clarify this point. A cheap calculator only carries out the particular operaions it has been 'hard-wired' to perform, whereas a computer permits various addresses (labels) to be introduced into the operations. Experience shows that this facility gives the computer enormous advantages over the simple calculator. We feel it must also be so for eucaryotic cells compared to procaryotic cells. The latter are 'hard-wired' to carry through a pre-set program, whereas the genes of eucaryotic cells can offer opportunities for the creation of new programs.

Viroids as Addresses

We can now give the concept of a viroid trigger a more precise meaning. The viroid is an address, or a system of addresses, that permits the host cell to search parts of the genome which are normally untranscribed.

It is well-known that most of the genome of eucaryotic cells does not give rise to gene expression. Instead of regarding it as redundant or parasitic, the unexpressed DNA can be thought of as sections of the genome for which the cell has lost, or has never had, appropriate addresses. Again in terms of a computer analogy, unexpressed DNA is a data bank which the cell has no means of addressing and so of using in its normal operation. This data bank contains, in our view, a cosmically determined store of evolutionary potential accumulated by cells throughout geological time.

Our hypothesis is that a viroid permits the cell to gain access to a portion of the genome that is usually outside its range of operation. This hypothesis is not without experimental support. Thus Keilin and Wang found that diseased legumes produced haemoglobin in certain cases[9]. This remarkable discovery demonstrated that the DNA of legumes contains the gene for haemoglobin, a gene which in plants normally goes unexpressed. The disease served as a trigger which permitted the plant to enter a normally forbidden part of its data bank.

Influenza Virus from the Genome

Our second hypothesis is that the addresses carried by an invading viroid set up a program modification in human cells (and similarly in cells of other animals) that generates the influenza virus from the human genome, a situation analogous to the generation of haemoglobin in legumes. This second hypothesis is also not without experimental support. Influenza is known to require the presence of an active cell nucleus before viral RNA or protein synthesis can take place[10,11]. Nor does the virus replicate in the presence of inhibitors of cellular DNA function such as actinomycin D or mitomycin C[12,13]. Moreover, endogenous viruses of Type C, which are isolated under certain pathological conditions, are known to be coded by the DNA of the host cell[14,15]. The viral genes have been discovered to be present in multiple copies in the cellular DNA, with over a hundred copies per haploid genome in some cases[15].

In 1976, Cox and Barry attempted to connect a part of the segmented genome of an avian influenza virus (fowl plague) with the DNA of a denatured chick cell[16]. The degree of homology obtained, 2 to 8 per cent, was not considered by these authors to warrant the conclusion that the virus could be obtained by transcription from the chick cell DNA. However, this work was done before it was realised that eucaryotic genes are split. Since a gene in separated bits cannot hybridise in an efficient simply-connected way with its connected product, this conclusion may need rediscussion. The fact that Cox and Barry obtained several per cent homology is already an interesting result, implying an association that could hardly be due to chance.

The Effect of Random Genetic Drift on Repeated Sequences

Sequences of normally untranscribed base-pairs of DNA are often repeated in the genome. We suppose the sequences responsible for abnormal transcription of the influenza virus happens to be repeated many times, very similar to the multiple copies found for Type C viral genes[15]. Because the several sequences are not involved in the usual operation of the cell, they are not protected by natural selection, and so drift randomly by copying-error mutations.

Random genetic drift does not lead to as disordered a situation as one might intuitively suppose. The situation is the following: If n is the number in the breeding group, most of the mutations acquired in a time span of more than $4n$ generations are shared either by all the members of the group or have been rejected. An appreciable fraction of the mutations acquired in the most recent $4n$ generations are variable, however, from one member of the group to another.

Antigenic Shift, Antigenic Drift, and Genetic Recombination of the Influenza Virus

From what has just been said, it follows that influenza virus transcribed from different sequences in the genome would show variations fixed in the population that increased more and more markedly as the generations increased above $4n$. A change of address from one long-existing sequence to another would therefore produce antigenic shift of the transcribed virus. We attribute shift to such a world-wide change of the addresses carried by incident viroids.

Antigenic drift can have two causes. Copies of sequences that are less than $4n$ generations old will differ more mildly one to another than longer-lived sequences. Person-to-person variability in the *same* sequence will also give the impression of genetic drift, as will partially distinct breeding groups. In this connection Nakajima et al.[17] have shown that isolates USSR/90/77 and USSR/92/77, which were evidently correlated both temporally and spatially, and which on the present theory would be expected to have been derived from the same sequence on the genome, nevertheless had base arrangements that differed by about one per cent. This is just of the order that we calculate for the degree of person-to-person variability due to stochastic drift, assuming an average breeding group over the past several million years of about 100,000.

Viroid invasions that supplied two or more addresses to sequences simultaneously would not only trigger the transcription of two or more forms of the virus but would create the possibility of recombinant events (of the kind that has recently been found[18-21]) occurring between the different transcribed forms. In such a complex situation the structure of the virus likely to multiply most would be decided by the immunological responses of the host.

The appearance of both H3 N2 and H1 N1 in the winter of 1977–78 can be explained by just such a multiple trigger. The address of the H1 N1 of 1977–78 happened to be the same as that of the H1 N1 of 1950. There is no need in the present theory to have recourse to the hypothesis that the H1 N1 of 1977–78 was a laboratory strain stored from 1950 that somehow effected a local escape and then spread itself around the whole world. Nakajima et al. remark that the H1 N1 of 1977–78 must have been frozen in nature for 27 years. So it was, we would say, on the human genome.

Primary Infection and Person-to-person Transmission

Once the virus has been transcribed, we suppose it to be capable of replication in the usual way. The virus can therefore spread from cell to cell, eventually causing a clinically observable attack of influenza. In principle, the virus could also spread from person-to-person but, because of the fragility of the free influenza virus, case-to-case transmission happens to be significantly weaker than the primary viroid infection.

If we suppose other viral respiratory diseases to be also activated by a viroid trigger, the ratio of the effectiveness of secondary transmission to primary infection must vary from one disease to another, depending on the persistence or otherwise of the free virus. For smallpox, the ratio happens to be high, for influenza it happens to be low.

Significance for Evolution

The ability of viroids to provide access to hitherto unaddressable parts of the genome could permit useful new genes to emerge, as well as producing the 'bad' results described above. The significance for evolution of switching new genes into the program of a cell could be very great, perhaps much greater than minor copying errors of the usual functional genes. One should not think therefore of invasion by viroids as being always 'bad'. Cells must take what they receive, some effects are 'bad', while others could permit the program of the cell to become increasingly sophisticated.

We suspect that many biologists will dislike the trend of our argument, because it suggests that evolution is driven by the larger universe outside the Earth. Our answer is that the facts of evolution suggest exactly such a picture. The usual closed-box-on-the-Earth picture of evolution appears quite incapable of explaining anything more than minor evolutionary features of biology. So far from the external picture being absurd or fantastic, what is surely absurd is to suppose that the hugely complex structure of microbiology, as for instance the detailed ordering of amino acids in 2000 or more enzymes, managed to evolve in an initially mysterious one-tenth or so of the Earth's history. The structure of microbiology has, in our view, a universal dimension that far transcends the minuscule size of our planet.

References

[1] Hoyle, F. and Wickramasinghe, N.C., *Diseases from Space* (J.M. Dent and Sons Limited, 1979).
[2] Irvine, W.M., Leschine, S.B. and Schloerb, F.P., *Nature* **283**, 748, 1980.
[3] Kates, M., Allison, A.C., Tyrell, D.A.J. and James, A.T., *Biochim. Biophys. Acta*, **52**, 455–466, 1961.
[4] Lodish, H.F. and Rothman, J.E., *Scientific American*, Jan. 1979, pp 38–53.
[5] Diener, T.O., *Science*, **205**, 859–865, 1979; Diener, T.O., Schneider, I.R. and Smith, D.R., *Virology*, **57**, 577.
[6] Edward, D.G.Ff., *The Lancet*, **2**, 664, 1941.
[7] Doolittle, W.F. and Sapienza, C., *Nature*, **284**, 601–607, 1980.
[8] Orgel, L.E. and Crick, F.H.C., *Nature*, **284**, 604–607, 1980.
[9] Keilin, D. and Wang, Y.L., *Nature*, **155**, 227, 1945; *see also* Ellfolk, N., *Endeavour*, **31**, 139, 1972.
[10] Follett, E.A.C., Pringle, C.R., Wunner, W.H. and Skehel, J.J., *Virol*, **13**, 394–399, 1974.
[11] Kelly, D.C., Avery, R.J. and Dimmock, N.J., *Virol*, **13**, 1155–1161, 1974.
[12] Barry, R.D., *Virology*, **24**, 398–405, 1964.
[13] Nayak, D.P. and Rasmussen, A.F.J., *Virology*, **30**, 673–683, 1966.
[14] Todaro, G.J. and Huenbar, R.J., *Proc.Natl.Acad.Sci.USA*, **69**, 1009, 1972.
[15] Todaro, G.J., Callahan, R., Sherr, C.J., Benveniste, R.E. and De Lareo, J.E., in *Persistent Viruses*, ed. T.O. Diener (Academic Press, 1978).
[16] Cox, N.J. and Barry, R.D., *Virology*, **69**, 304–313, 1976.
[17] Nakajima, K., Desselberger, U. and Palese, P., Nature, **274**, 334–339, 1978.
[18] Desselberger, U., Nakamima, K., Alfino, P., Pederson, F.S., Heseltine, W.A., Hannoun, C. and Palese, P., *Proc. Natl. Acad. Sci. USA*, **75**, 3341–3345, 1978.
[19] Young, J.F. and Palese, P., *Proc. Natl. Acad. Sci. USA*, **76**, 6547–6551, 1979.
[20] Bean, W.J., Cox, N.J. and Kendal, A.P., *Nature*, **284**, 638–640, 1980.
[21] Scholtissek, C., Rohde, W., von Hoyningen, V. and Rott, R., *Virology*, **87**, 13–20, 1978.

5

INFLUENZA IN SOME WEST-COUNTRY SCHOOLS (1978)

The distinguished epidemiologist Charles Creighton maintained, as late as the final decade of the 19th century, that influenza is not a transmissible disease. In his book 'History of Epidemics in Britain'(Cambridge University Press, 1891), he discusses the influenza epidemics of 1833, 1837, and 1847, in which medical opinion held that populations living over considerable areas were affected almost simultaneously. This evidence suggested to Creighton a 'miasma' over the land rather than a disease which must spread itself from person to person. If one substitutes for 'miasma' the phrase 'viral invasion from space', one has a similar position to that which we proposed in a recent article (*New Scientist*, 17 November 1977).

Creighton's position has not been taken seriously in the present century, because as M. W. Kaplan and R. G. Webster (*Scientific American*, December 1977) remark:

"By the end of the 19th century . . . . the microbiological concept of infectious disease had taken firm root."

Strictly, the microbiological concept requires only that victims of the disease should acquire the influenza virus from outside of themselves, which of course they would do if the invasion came from space. But such an idea seemed so much less plausible to medical opinion than the concept of person-to-person transmission that person-to-person transmission was not even considered as a hypothesis to be tested. It became an axiom.

Yet the evidence for Creighton's position did not go away. It became stronger as the following example, a quotation from a Sardinian doctor commenting on a new variant of influenza in 1948 shows:

"We were able to verify . . . . the appearance of influenza in shepherds who were living for a long time alone, in solitary open country far from any inhabited centre; this occurred absolutely contemporaneously with the appearance of influenza in the nearest inhabited centres."

Accompanying this seemingly lightning contageousness of influenza, there is acutely puzzling evidence for the failure of the virus to travel very short distances except in quite long periods of time:

"It (influenza) was present for the first time at Joliet *four weeks* (our italics) after it was first detected in Chicago, the distance between those areas being only 38 miles . . . . "
(L. Wienstein, Influenza–1918, a Revisit, *New England Journal of Medicine*, 6 May 1976).

It is also an unsolved puzzle for the axiom of person-to-person transmission that modern air travel has not had much effect on the rate of spread of epidemic outbreaks of influenza. For instance, in 1968 there was a very long 12 week interval between the incidence of such outbreaks in California and Florida, despite the considerable volume of daily air traffic between major cities in these two states.

It is often said that clear evidence for the person-to-person transmission of influenza can be found in epidemic outbreaks within special communities – military bases, schools, isolated island people. However, it often happens in this data that the number of victims rises very steeply, with an e-folding time of about 1 day, much less than the 2–3 days incubation period of the disease itself. (Numbers on successive days might look like 3, 7, 20, 70, 240). Such very steep rises are possible on a person-to-person basis only if each afflicted person passes the disease to several other persons, the disease 'cycle' being strongly supercritical. It is then a requirement that the outbreak cannot burn itself out until essentially all susceptible persons in the group have succumbed to it. Yet such steeply rising epidemic outbreaks often end with only some 25 percent of the group affected, whereas antibody titres for the virus in question suggest that the susceptible percentage is much larger than this.

The best example known to us for the existence of person-to-person transmission occurred on the island of Tristan da Cunha in the winter of 1971, the virus being A/Hong Kong/68, (J. Mantle and D. A. J. Tyrell, *J. Hyg., Camb.* (1973), *71*, 89).

The steep incidence in this case was consistent with a very high fraction of the islanders being affected by the disease, 96 percent. Yet even for this example it is something of a puzzle to understand how an apparently exceedingly susceptible population managed to escape infection by A/Hong Kong/68 from all the ships which called at the island between 1968 and 1971 (fishing vessels put in there quite frequently for supplies).

With all these difficulties and questions in mind, and with a new influenza pandemic in our midst, it occurred to us to attempt a test of the concept of person-to-person transmission from an assessment of its effect on school absenteeism. For this purpose, we examined attendance records at a number of west-country schools, mostly in the Cardiff area. Our object was to determine, for each individual school, the time dependence of absenteeism due to influenza during the epidemic. The entire school-going population would have been susceptible to at least one of the influenza sub-types (H1 N1) which is known to be involved in the present epidemic. Any excess of absences extending across an entire school (above a well-defined normal average value) may be used as a measure of the incidence of influenza on a given day. Such data has the disadvantage that the absence of pupils was not always medically confirmed as due to influenza. There was considerable circumstantial evidence, however, that influenza was in fact the overwhelming cause of the above-average pupil absences recorded by the schools over the time intervals in question. First, we were careful only to choose those schools which otherwise have high percentage attendance records. (Unfortunately, this requirement ruled out most city-centre schools where truancy is endemic). Second, the epidemic outbreaks in the schools occurred from mid-January to mid-February when medically-attested outbreaks of influenza were occurring in the general Cardiff area. Third, the symptoms subjectively reported by pupils were those characteristic of influenza–headache, fever, blockage of the upper respiratory tract. Fourth, a portion of our data covered boarders in public schools, and in these cases school matrons were in no doubt that the disease was in fact influenza.

Data obtained from school attendance records have a considerable advantage over medically-attested data and over industrial sick-leave data. The latter are necessarily concerned with the minority of severe cases, whereas absence from school has a lower decision-level that includes much milder cases. School attendance records should therefore give a rather complete picture of the incidence of the disease.

The schools we have consulted (with the exception of Atlantic College, which we shall discuss later) are set out in Table 5.1, together with numbers of pupils, locations, and approximate catchment areas for day-pupils. The location of the schools and catchment areas where they are fairly accurately known are shown in Map 1. In the following figures (Figures 5.1–5.3) the ordinate represents percentage absence in excess of average absence, the average absence being estimated from attendance records over a period adjacent to, but well clear of, the epidemic.

Figure 5.1 shows percentages for young pupils in the 1st and 2nd forms at Howell's, compared with percentages for the whole school. We interpret this lack of substantial variation of the attack-rate with respect to age as implying a more or less constant measure of inherent susceptibility to the disease for young persons taken over the whole Cardiff area. Hence we think that the very marked variations from school to school shown in Figures 5.2 and 5.3 cannot be attributed to inherent differences among the pupils themselves. The differences of the curves in Figures 5.2 and 5.3 must be attributed, we believe, to differences in the geographical locations of the schools and differences in their catchment areas.

Although the maximum separation of the schools represented here is only about 20 miles, this distance is not small compared with the average height of the tropopause, 8–10 miles. Significant fluctuations of atmospheric pressure and density might be expected to occur over horizontal distances which are comparable to 10 miles, and such fluctuations could in turn produce variations in the time of descent of micrometeoritic dust. Smaller-scale fluctuations could also be

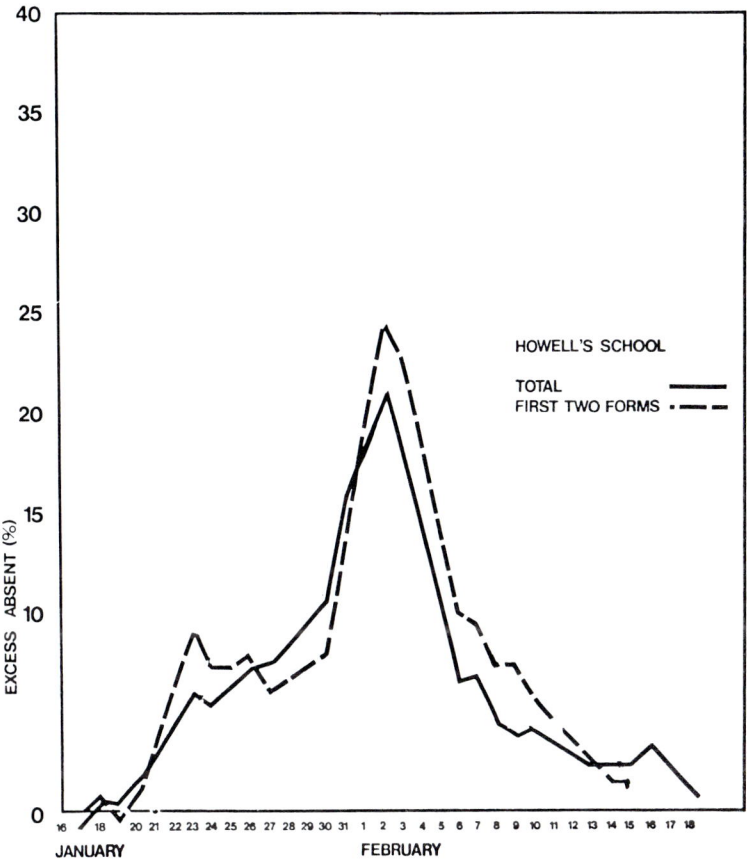

Figure 5.1.
Excess absences (percent) above normal for Howell's School. Three curves show absences for whole school, boarders only and first two forms only.

superposed on these natural meteorological conditions. The effects of urban activity could result in updrafts of air over cities which might tend to shield city cores from the settling of micrometeoritic matter. On an even finer scale, we may expect individual houses or clumps of houses to provide local protective blankets of rising air.

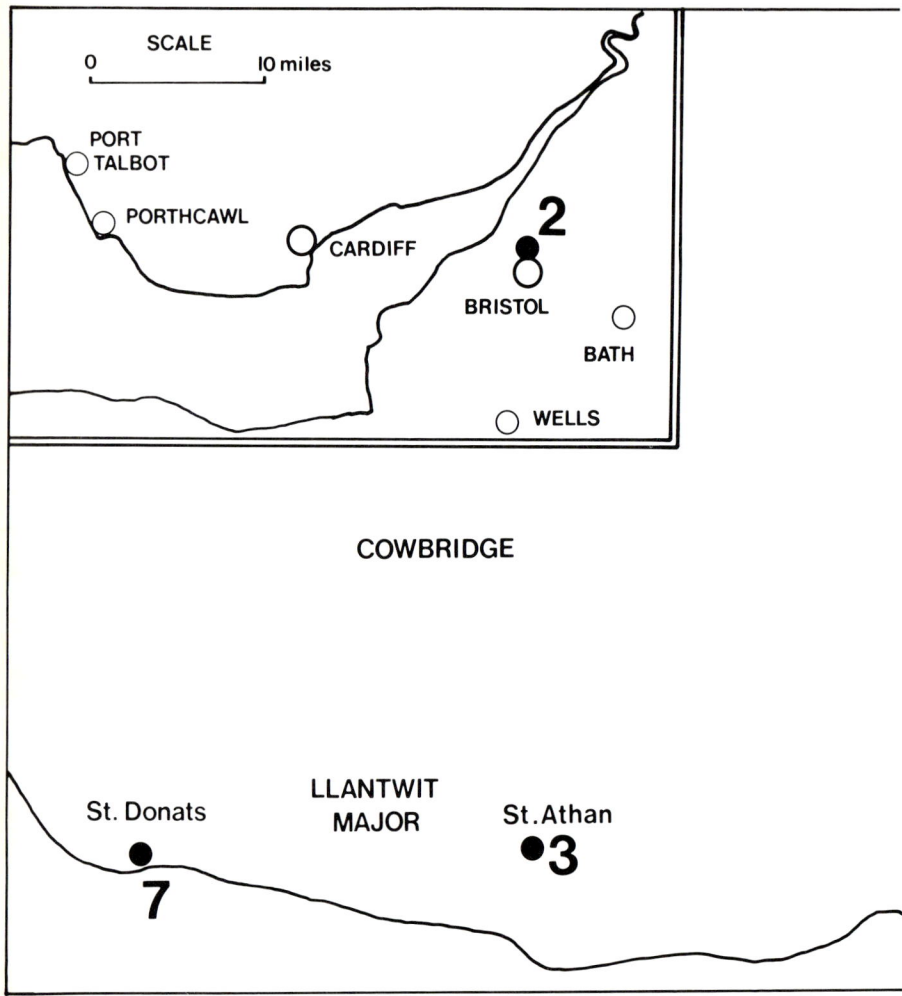

Map 1.
Map of Cardiff area and west-country (inset) showing schools studied. Numbers refer to Howell's (1), Clifton (2), Balfour House (3), St. Cyres (4), Llanishen High (5), Cardiff High (6) and Atlantic College (7). Hatched areas show catchment areas (where known) around schools.

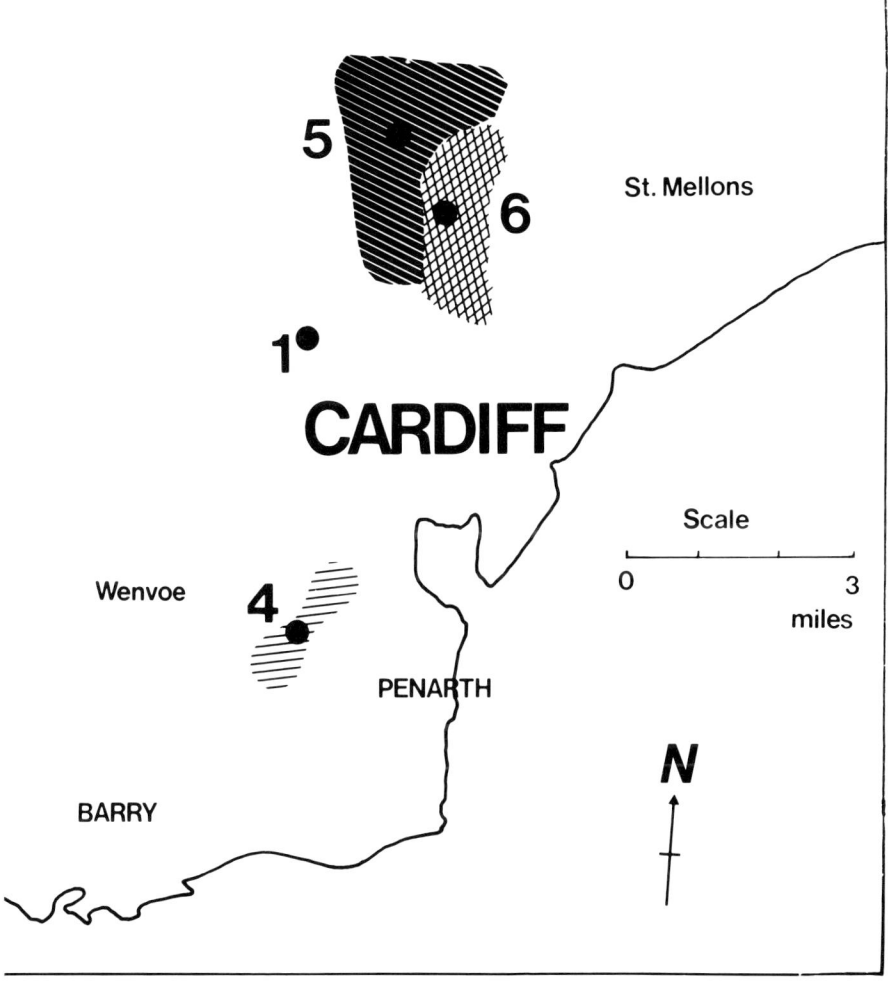

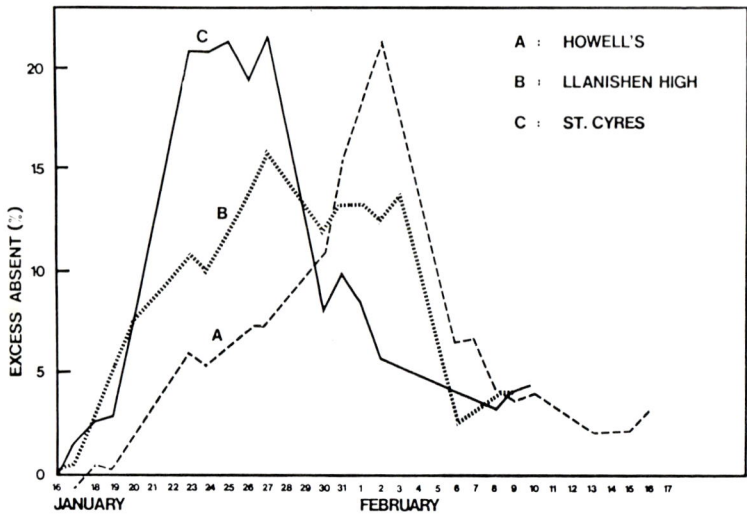

Figure 5.2.
Excess total of absences (percent) above normal for Howell's Llanishen High and St. Cyres.

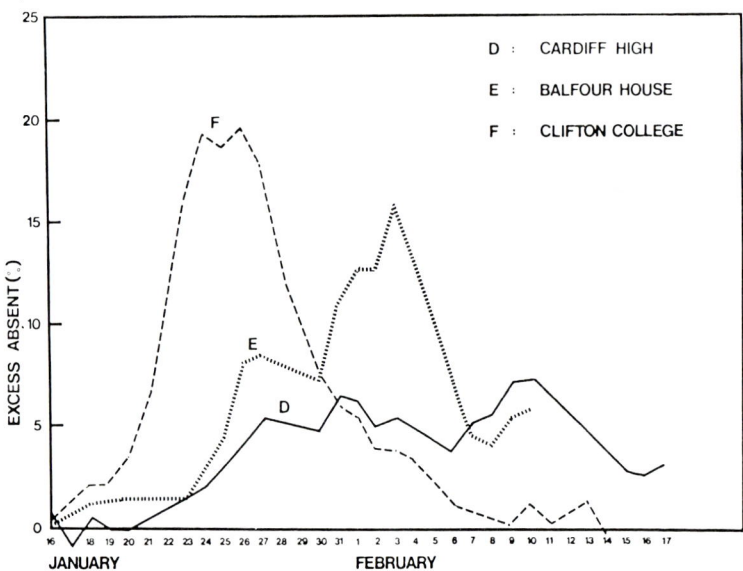

Figure 5.3.
Excess total absences (percent) above normal for Cardiff High, Balfour House and Clifton College.

We do not think that such big differences as those we have found can be explained by arguing that physical variations in the schools themselves were responsible for fluctuations in the mode of person-to-person transmission. Class numbers in all cases range from about 25 to 35, while desk-spacings between pupils are quite standardised – both class numbers and spacings being essentially determined by economic considerations that are common to all schools. Nor do we think there is much difference in the play-habits of pupils in the different schools, as for instance in the extent to which they whisper, talk, or huddle together. In short, we see no satisfactory way to interpret the wide diversity of incidence curves in Figures 5.2 and 5.3 in terms of person-to-person transmission. A more plausible explanation, in our view, is that such variations are caused by fluctuations in the time of descent of virus-bearing particles.

It has often been speculated that restricted conditions in dormitories may largely be responsible for epidemics in military barracks and public schools. Yet in two of the schools in our sample, there was remarkable evidence against the importance of person-to-person transmission within dormitories.

We first discuss a most curious situation which occurred for Atlantic College at St. Donats. This sixth form school has many unique features. It is built around the 14th century St. Donats Castle and is situated on an exposed cliff overlooking St. Donats Bay (see map 1). The nearest town is Llantwit Major. The school draws students from all parts of the world who wish to qualify for admission to British and American universities. It is financed by international educational organisations and has excellent recreational, residential and academic facilities. There are 343 students between the ages of 16 and 19, all in residence, in 8 houses scattered on the campus. During the period 13 February–3 March, the school suffered an attack of acute upper respiratory disease. No class attendance records are kept, but the Sister in charge of the Medical Centre gave us access to her records of case admissions to the school infirmary. All 48 recorded cases were found to be febrile, the duration of acute illness being 4–5 days. No illness was recorded during the period in late January and early February when peak absenteeism was found both in the Cardiff and Bristol areas.

Dormitory facilities consist of 17 rooms with 3 students to a room, 70 rooms with 4 students each, and 2 rooms with 6 students each. If the influenza virus was picked up by the 48 'victims' during daytime activities unconnected with the dormitories, *and if there were no person-to-person transmission within the dormitories*, the 48 cases would be distributed randomly with respect to the sleeping facilities. The distribution would then be similar to the random allotment of 48 objects into 89 boxes (with the neglect of slight variations in the occupancy of a small fraction of the rooms). In such a distribution, the expected ratio of boxes with 1 object allotted to boxes with 2 objects allotted to boxes with 3 objects allotted is given by the ratios of binomial terms $1 : 1/2 \times 47/88 : 1/6 \times 47.46/88^2$. Thus a typical result in such a random allotment would be 31 boxes with 1 object, 7 boxes with 2 objects, 1 box with 3 objects. At Atlantic College there were 35 dormitories with 1 victim, 5 dormitories with 2 victims, and 1 dormitory with 3 victims. The clear implication is that the conditions necessary for a random situation existed, person-to-person transmission within dormitories being a negligible factor.

At Howell's, person-to-person transmission within dormitories was also a negligible factor, but unlike Atlantic College, the geographical location of the dormitories was important, implying that a substantial fraction of cases were actually contracted within the dormitories themselves. Boarders at Howell's occupy 4 houses which fall into two distinct pairs (see map 2). The two members of a pair are geographically very close, but the two pairs themselves are distinctly separated. The numbers of victims for the two pairs were (4, 5) and (12, 14), the attack rate in the former pair being significantly less than in the latter.* The distribution of cases

---

*This otherwise uncanny difference is readily explained in our view as a consequence of a ground-level viral infall which is patchy on the scale of 100 meters.

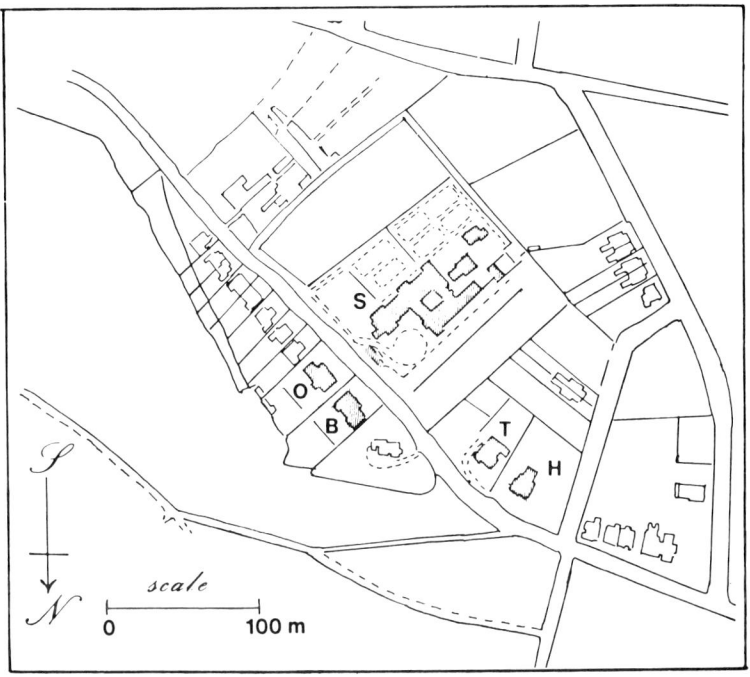

Map 2.
Map of area around Howell's school. S refers to Main School; O refers to Oaklands House; B refers to Bryn Taff House; T refers to Taylor House; H refers to Hazelwood House.

with respect to the individual dormitories of each house is shown schematically in Figure 5.4, in which each victim has been represented by a number, a number which gives order in a sequence related to the day of outbreak.

Thus 16 means the 16th victim to report the onset of illness. ('H' refers to a weekend boarder who was taken ill at home approximately at the same time as the bulk of the boarders). It will be seen that when two or more victims occupy the same

Table 5.2

Hazelwood and Taylor (Howell's School)

These two houses contain 51 boarders distributed in rooms containing 1–6 beds each. We estimate here the expected distribution of 26 victims to be compared with actual numbers.

No. of beds in room (n)	No. of rooms	Fraction of boarders in rooms with n beds (f)	Numbers of victims expected for random distribution 26f	actual numbers
6	3	18/51	9	10
5	2	10/51	5	4
4	2	8/51	4	5
3	2	6/51	3	3
2	3	6/51	3	1
1	3	3/51	2	3

dormitory, a close correlation of the order numbers frequently exists, implying an essentially simultaneous onset of the disease. Such associations are not due therefore to one victim passing the disease to the other. With such clearly correlated victims taken account of, the remaining examples of multiple outbreaks in particular dormitories are again in accordance with a random allottment of cases. (see Table 5.2)

From the limited sample of west-country schools we have studied, it has become clear that Dr. Creighton's position on the non-transmissible nature of epidemic influenza might at last be vindicated. The indications are that during epidemics (such as the one currently raging) the influenza virus descends through the atmosphere and settles with an exceedingly fine-scale patchiness at ground level. We suspect that a great deal of useful, even decisive, information probably lies hidden in the attendance records of schools throughout the country,

INFLUENZA IN SOME WEST-COUNTRY SCHOOLS (1978)

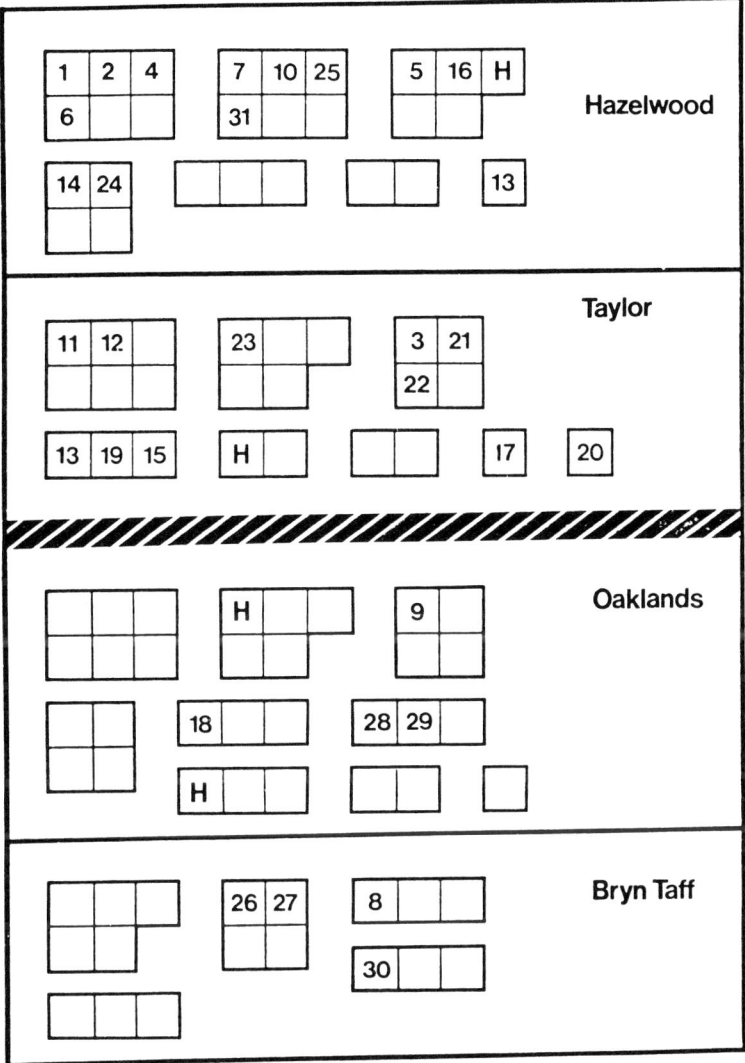

Figure 5.4.
Schematic plan of dormitories at Howell's for four Houses: Hazelwood, Taylor, Oaklands and Bryn Taff. Boxes represent rooms, partitions within boxes are occupied beds. Numbers show order of victims reporting to the school nurse. H stands for a week-day boarder who had influenza over the weekend.

especially since records for the pandemic years of 1957 and 1967 are presumably still available. It will, of course, need some co-ordinated programme to bring out the full weight of this evidence, but we think (and this a main reason for the writing of this article) that the effort needed to mount such a programme could be well rewarded.

We are grateful for the assistance and co-operation of headmasters, headmistresses, nurses and matrons at schools. We also acknowledge advice from Drs. M. S. Pereira, P. Mann and C. W. L. Howells of the Public Health Laboratory Services. Finally, we thank Harry Abadi and Priya Wickramasinghe for assistance in securing data.

6

DOES EPIDEMIC DISEASE COME FROM SPACE? (1977)

Comets have been regarded with awe and trepidation by many ancient cultures in widely separated parts of the globe. Almost invariably they have been regarded as bad omens–bringers of pestilence and death. In this article we argue that a cometary impact on the Earth could have led to the start of terrestrial life. Even today, the periodic influx of cometary debris in the form of micrometeorites may be responsible for waves of disease which sweep the planet.

The usually accepted theory of the origin of life is due to A.I. Oparin and J.B.S. Haldane. This theory involves several uncertain assumptions concerning conditions on the primitive Earth. The conversion of simple inorganic molecules such as water, methane, ammonia into a "primeval soup" of prebiotic molecules requires a high frequency of energising events such as thunderstorms and lightning. And the stability of prebiotic molecules so formed presupposes the presence of an atmosphere which is reducing rather than oxidising–it must tend to remove rather than add oxygen as it does today. A large initial excess of molecular hydrogen is usually assumed–an assumption for which there is no astronomical or geological basis. Indeed, an original oxidising atmosphere appears more likely, and in this case no primeval soup could have developed.

Laboratory experiments, notably by Stanley Miller, Cyril Ponnamperuma and their colleagues, have shown that molecules such as hydrogen, water, methane and ammonia can be converted into amino acids, nucleic acid bases and sugars under controlled conditions. But these demonstrations do not constitute proof of the Oparin-Haldane theory, as is usually claimed. Firstly, primitive Earth conditions may not have borne any similarity whatsoever to those used in the experiments. Secondly, even if pre-biotic molecules do form by processes of this type, their concentrations in primitive lakes and oceans would most probably have been too low to lead to

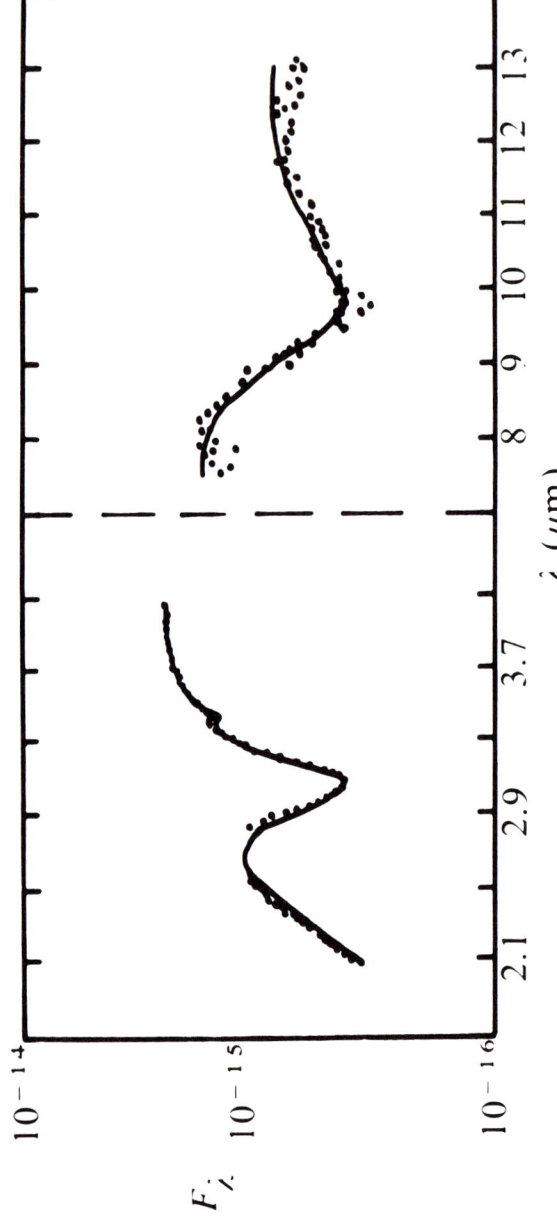

Figure 6.1.
The infrared spectrum (right) of a polysaccharide 'dust' model compared (left) with that of the BN source of the Orion Nebula.

the start of life. Furthermore, there is a disconcerting lack of evidence for any large-scale nitrogenous carbonaceous deposits in the oldest sedimentary rocks. Such deposits would indeed be expected to occur if the primeval "soup" existed for a long enough period; and their absence in the geological record may be construed as evidence against the soup.

We have recently developed the point of view that the essential building blocks of life – amino acids, nitrogen-bearing heterocyclic compounds and polysaccharides – are formed in space. These compounds occur in large quantities throughout the Galaxy. Figure 6.1 shows the exceedingly close agreement between the predictions of a polysaccharide "dust" model and the observed infrared radiation from the well-known BN source in the Orion Nebula. Prebiotic molecules of these sorts could be mopped up by cometary-type objects and injected into planets such as the Earth.

Comets appear to be a highly promising venue for the actual origin of life. Typically a few kilometres in radius, cometary nuclei are the source of all the observed cometary phenomena – including molecular comas, plasma and dust tails (see Figure 6.2). Although the recent long-period comet Kohoutek was a visual disappointment, there were important observations here of organic molecules such as hydrogen cyanide and methyl cyanide. Smaller molecules and radicals such as CO, CO_2, N_2, C_2, C_3, OH, CH, CH_2, NH, NH_2, CN and H_2O which are present in cometary comas and tails could be degradation products of complex biochemicals and of H_2O. There is also an infrared emission feature at 10 micrometres in dust tails which is possibly due to polyformaldehyde, polysaccharides, or a similar organic polymer.

Comets most probably originated within the gas cloud from which planets formed, condensing as a by-product of the formation of the outer planets Uranus and Neptune. Some thousand Earth masses of hydrogen and helium expelled from the primitive Sun during its early over-luminous phase were carried to the distances of the present outer planets, along with a fine smoke of ices and silicates which condensed further in. Such a disc of solar-system material will mop up a mass

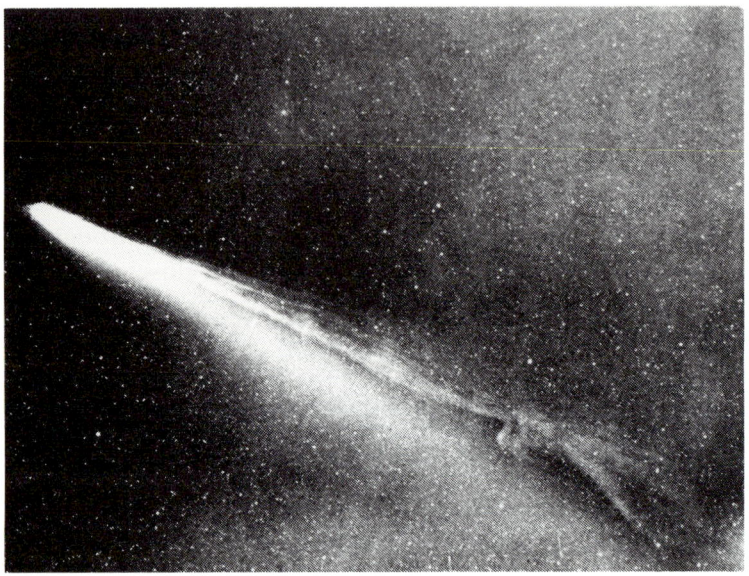

Figure 6.2.
Photograph of Comet Kohoutek (1973).

comparable with its own mass of interstellar material from the dense interstellar cloud in which it is immersed due to random oscillations within it over a timescale of a hundred million years. The frictional heating thus generated will lead to evaporation and loss of hydrogen and helium from the Solar System, but not of the heavier organic molecules. The mass of prebiotic material added to the disc is likely to have been about 10 Earth masses and this material would condense on pre-existing nuclei composed mainly of water-ice. The fraction by mass of cometary material in the form of prebiotic molecules may be estimated at about 30 per cent – a concentration that could not be matched in any later terrestrial situation.

The aggregation of the outer planets may be estimated to have taken about 300 million years – considerably longer than the aggregation time for the terrestrial planets Mercury, Mars,

Earth and Venus. During this period many cometary-type bodies which formed at the distances of the outer planets would have been randomly deflected into elongated orbits which crossed the orbits of the inner planets, and several direct encounters with the Earth would have occurred. An initial "hard" encounter would have led to the formation of a gaseous terrestrial atmosphere. Subsequent encounters would have produced "soft" landings, in which complex biochemical structures were preserved.

During close perihelion passages of comet-type bodies, volatile material (particularly water) from the nucleus would tend to mingle with the dust and organic molecules that were condensed upon the nucleus. This would produce high aqueous concentrations of interstellar prebiotic molecules including polysaccharides, amino acids, and nitrogen-bearing heterocyclic compounds, underneath a frozen icy outer skin. The temperature of the cometary nucleus would oscillate between 300°K at perihelion and 100°K at aphelion distances of the order of the radii of the orbits of outer planets. Such oscillations, which produce periodic episodes of melting, evaporation and re-freezing, provide a selective mechanism for the emergence of those molecular structures which can best withstand break-up during the hottest phases and which could also preserve themselves during episodes of intense cold. Prebiotic material could therefore undergo intermittent chemical processing and evolution, in a process lasting many millions of years. Ultraviolet sunlight incident on melted cometary ices could induce polymerisation reactions, leading to the elaboration of prebiotic molecules into more complex structures such as polynucleotides, polypeptides and porphyrins. And such macromolecules, produced in one cycle, could hold "survival patterns" in a frozen state until the next perihelion passage when simpler molecules could be recast into "predetermined" structures.

The first living organisms may have evolved in response to these selective pressures. Photosynthetic bacteria, able to oxidise hydrogen sulphide anaerobically, may have been the result of this evolution. We know that such organisms are viable even where the sunlight intensity is as low as one-tenth of that required for other photosynthetic bacteria to survive. This property would be an asset for primitive cometary organisms, which must attempt to maximise their "effective life-span" by operating under conditions of the lowest possible sunlight intensity compatible with immersion in liquid water. The requirement for survival at perihelion, when the temperature is high, favours the evolution of thermophilic bacteria–similar to those found in certain hot springs on Earth where the temperature can be as high as 95°C. Terrestrial life could well have originated about four billion years ago by the soft landing of an icy comet already containing primitive organisms.

The number of close perihelion passages that a comet can survive before becoming completely stripped of volatiles is probably a few thousand. Carbonaceous chondrites could represent materials from comets denuded of volatiles, but retaining a residue of silicates and more resistant organic polymers. In 1961 G. Claus and B. Nagy reported that they had discovered in certain meteorites organised microscopic structures resembling bacterial fossils. Later, the same meteorites were found to contain a host of biological molecules including amino acids, porphyrins and nucleic acid bases. Although some of the structures observed by Claus and Nagy later proved to be pollen grains, it remains possible that others are genuine microbiological fossils which owe their origin to primitive cometary organisms of the type discussed here.

Recent studies by R.S. Rajan, D.E. Brownlee, D. Tomandle, P.W. Hodge, I.H. Farrar and R.A. Britten, of cometary debris in the form of micrometeorites in the stratosphere have shown chemical similarities to Type 1 carbonaceous chondrites. If a cometary impact led to the start of life, the question arises: would subsequent arrivals of cometary material carry biological or prebiological material which might affect terrestrial biology? The boldest answer must be yes; that is to say, extraterrestrial biological invasions never stopped and continue today. These invasions could take the form of new viral and bacterial infections that strike our planet at irregular intervals, drifting down onto the surface in the form of clumps of meteoritic material probably similar to those studied by Dr Rajan and his colleagues. Recurrent waves of new disease could represent renewed attempts at the evolution of life on comets, infection reaching the Earth when its orbit crosses the trails of debris from these comets.

Reports of the sudden spread of plagues and pestilences punctuate the history of many countries. The most recent such disaster was the 1918–19 influenza pandemic in which 30 million people died. Different epidemics, scattered throughout history, bear little or no resemblance one to another. But they all share a common property of afflicting entire cities, countries or even widely separated parts of the Earth in a matter of days or weeks. Thucydides describes the plague of Athens in 429 BC thus:

"It is said to have begun in that part of Ethiopia above Egypt... On the city of Athens it fell suddenly, and first attacked the men in Piraeus; so that it was even reported by them that the Peloponnesians had thrown poison into the cisterns."

A similar description of sudden onset and rapid global spread is relevant to almost all earlier as well as later epidemics. Such swiftness of transmission is hard to understand if, as is usually supposed, infection can pass only from person to person or be carried by vectors such as lice and ticks. And this explanation is particularly untenable for widespread epidemics which occurred before the advent of air travel, when movement of people across the Earth was a slow and tedious process.

The general belief, which is not well proven, is that major pandemics, such as influenza in recent times, start by random mutation, or genetic recombination, of a virus or bacterium which then spreads by direct person-to-person contact. If this is so, it is somewhat surprising that major pandemics are relatively short-lived, usually lasting about a year, and that they do not eventually affect the entire population, which would not have a specific immunity to the new organism. We contend that primary cometary dust infection is the most lethal, and that secondary person-to-person transmissions have a progressively reduced virulence, so resulting in a declining incidence of the disease over a limited period. Primary infections of a human population could occur directly by contact with infected meteoritic dust, or indirectly by meteoritic infection passing to other creatures such as mosquitoes, rats and lice which act as intermediaries.

The abrupt appearance in the literature of references to particular diseases is also significant in that they probably indicate times of specific invasions. Thus, the first clear description of a disease resembling influenza was in the 17th century AD, while the earliest reference to the common cold was in about the 15th century AD. Also it is significant that earlier plagues, such as that in Athens, do not have easily recognisable modern counterparts.

Major epidemics of disease could be caused when the Earth crosses the debris of new long-period comets. Relatively minor variants of the "same" disease – eg. the common cold – could be due to more frequent, regular passages of the Earth through debris of shorter-period comets.

The factors governing the actual pattern of global incidence for any particular extraterrestrial invasion could be complex. If bacteria or viruses are dispersed in a diffuse cloud of small particles, the incidence of disease may well be global. On the other hand, a smaller disintegrating aggregate of infective grain clumps falling over a limited area of the Earth's surface could provide a geographically more localised invasion. Systematic effects such as air currents over the Earth's surface could also be relevant in controlling the transport and dispersal of particles. In particular, certain latitude belts might well be more favoured than others for either the accumulation and settling of these particles, or their avoidance. Furthermore, spatial variations in settling times, corresponding to variations in atmospheric conditions at different locations, could mimic a situation where an epidemic apparently spreads from a localised focus – the spread having no casual connection whatever with the terrestrial "focus".

Our suggestion, if correct, would have profound biological, medical and sociological implications. A continual microbiological vigil of the stratosphere may well be necessary to eliminate the havoc which will ensue from extraterrestrial invasions in the future.

7

THE INFLUENZA EPIDEMIC OF 1977/78

With a view to testing the cosmic theory of life we studied the incidence of influenza in the winter of 1977/78 as it affected school children in England and Wales. We obtained data connected with this epidemic by circulating a questionnaire to independent schools in England and Wales. Before using this data in our publication 'Diseases from Space' we sent the following extended letter to Headteachers to seek permission for naming their particular schools:

Subject: Distribution of Influenza Cases According to School Houses

Dear Headmaster/Headmistress,

A number of schools, including your own, have sent us interesting comments and data in reply to our recent questionaire. While it is not our intention to report this information publicly (in association with the names of schools) without first seeking permission to do so, we think you will not mind a restricted circulation being made now among the other schools in this particular group (Table 7.1): We think that each school may be interested to know what happened to the others! If indeed you would be willing for your data to appear more widely than this, we would be grateful if you would give us the permission to include present references to your school in our eventual report on this survey.

According to our view, influenza cases are contracted through the breathing of a virus that falls down through the air, and not through person-to-person transmission. The incidence of the virus will always be patchy, both for broad meteorological reasons and because irregularities at ground level (e.g. buildings) cause local eddies. The patches can vary both in their concentration of the virus and in their physical scale. The time intervals over which different locations are

Table 7.1

List of Schools

Berkhamsted School, Berkhamsted
Eton College, Windsor
Giggleswick, Settle
Headington, Oxford
King Edward's Witley
Oakham, Rutland
Queen Margaret's, York
Queenswood, Hatfield
Rydal, Colwyn Bay
St. Audrie's, West Quantoxhead
St. John's Leatherhead
St. Mary's Calne
St. Michael's, Oxted
Seaford College, Petworth
Shiplake, Henley-on-Thames
Stoke College, Stoke-by-Clare
Stowe, Buckingham
Sutton Valence, Maidstone
Upholland College, Skelmersdale
Ursuline Convent, Westgate-on-Sea

exposed to the virus can also be variable. Some locations may unfortunately be hit by a particularly intense patch, or sequentially hit by several such patches, while other more fortunate locations may scarcely be affected at all. The full range of expected possibilities is therefore wide and the consequences complex. We shall be concerned in this paper with attempting to unravel the intricacies of the situation by examining the association of influenza victims with school houses, and in particular by attempting to decide for individual schools whether the distribution of victims in houses should be considered random or not. The issue of randomness (or otherwise) is important because it establishes the following criteria for the incidence of the influenza virus:

Conditions leading to a random distribution of victims in school houses

The physical scale of patchiness of the virus *greater* than the environs of the school,

or

The incidence of the virus only at times when pupils are away from their houses (e.g. during general school hours).

Conditions leading to a non-random distribution of victims in school houses

The physical scale of patchiness of the virus *less* than the environs of the school,

and

the virus, or a portion of it, incident at times when pupils are in their houses.

Diversion on Statistical Considerations

Write N for the total number of pupils in a school, and M for the total number of influenza victims in an epidemic. The attack-rate p for the whole school is simply M/N, and the average random allotment of cases to a house with n pupils is np.

It is to be observed, however, that if a lottery of M objects, (e.g. apples) equally weighted for every pupil, were made, the number that would be found to go into the house with n pupils would hardly ever be precisely *np*. But if such an allotment of M objects were made every day over an extended time interval, the number going each day into the house with n pupils would be found to fluctuate around *np*, and the *average* allocation would turn out to be *np*.

The fluctuation from np on a typical day would be about $\sqrt{np(1-p)}$, the so-called standard deviation. Occasionally, however, fluctuations larger than this would occur. On about one occasion in twenty the fluctuation would be as large as $2\sqrt{np(1-p)}$, with a rapidly decreasing frequency of occurrence for still larger fluctuations. Thus a fluctuation as large as $3\sqrt{np(1-p)}$ would happen only very rarely, on somewhat less than one occasion in 300.

When one has no *a priori* knowledge that a distribution is actually random, when indeed it is the aim to decide whether or not a distribution is random, the usual procedure is to take a look at the fluctuations. If the fluctuations are no greater than $\sqrt{np(1-p)}$, one would not query the possible randomness of the distribution. At $2\sqrt{np(1-p)}$ one begins to hesitate, with the distribution looking as if it might be non-random. And at $3\sqrt{np(1-p)}$ or more one can assert, with little chance of being mistaken, that the distribution is non-random. We shall adopt this method of approach in our discussion of the distribution of influenza cases into school houses, beginning with schools where the distribution was quite possibly random, and then passing to further schools where the distribution was almost surely non-random.

Schools with Possibly Random Distribution of Influenza Cases into Houses

St. John's, Leatherhead

Treating boarders and day pupils at St. John's together, since the attack-rate p was the same (64 per cent) for boarders and day pupils, we have the following analysis of the distribution with respect to houses.

$N = 446$, $M = 285$, $p = M/N = 0.64$

House	n	np	Actual no. of cases	Fluctuations $\div \sqrt{np(1-p)}$
South	65	41.54	42	0.12
East	64	40.90	34	1.80
West	63	40.26	39	0.33
Churchill	67	42.81	47	1.07
North	61	38.98	36	0.79
Surrey	64	40.90	48	1.85
Montgomery	62	39.62	39	0.16

The fluctuations are largest for East and Surrey, but they are not yet outside the range to be expected for a random distribution.

Of the possible criteria given above for a random distribution of victims into houses, an attack of the virus during school hours is the more likely, since such an attack would hit boarders and day pupils equally. The columns of the questionnaire showing absences from classes indicate that the attack occurred on or around Thursday, 9th February.

The experience at St. John's has the added interest of demonstrating that the susceptibilities of boarders and day pupils are very closely the same.

Queen Margaret's, York

The data from Queen Margaret's permits an analysis with respect both to houses and to the six upper forms of the school. The comparison is interesting, as will be seen from the following results:

$N = 247$, $M = 143$, $p = 0.58$

House	n	np	Actual no. of cases	Fluctuation $\div \sqrt{np(1-p)}$
Red	42	24.32	20	1.35
School	38	22.00	23	0.33
Hall	38	22.00	15	2.30
Garry	37	21.42	23	0.53
St. Aidan's	32	18.53	25	2.32
Dower	61	35.32	37	0.44

Form	n	np	Estimated* no. of cases	Fluctuation $\div \sqrt{np(1-p)}$
VI	about 33	19.14	17	0.75
VA	about 33	19.14	31	4.18
VB	about 33	19.14	9	3.58
IVR	about 33	19.14	27	2.77
IVA	about 33	19.14	19	0.05
IVB	about 33	19.14	13	2.17

* Estimated on the assumption that the average length of absence of a victim from class was the same for all forms.

While one might perhaps assign a measure of non-randomness to the distribution with respect to houses, there is clear non-randomness in the distribution with respect to forms. As at St. John's, Leatherhead, pupils at Queen Margaret's were clearly infected during general school hours, and moreover in a patchy way with respect to different parts of the school buildings. Class absences indicate that the main infection occurred around 2nd and 3rd February.

Seaford College, Petworth

The data from Seaford College leads to the following analysis:

$N = 440$, $M = 73$, $p = 0.17$

House	n	np	Actual no. of cases	Fluctuation $\div \sqrt{np(1-p)}$
Co	104	17.25	21	0.99
M	86	14.27	13	0.37
Ch	82	13.60	17	1.01
Ad	86	14.27	14	0.08
Ky	82	13.60	8	1.66

The fluctuations are evidently not large enough for a non-random situation with respect to houses to be indicated.

Accepting a random distribution, we can ask: which of the alternative conditions leading to randomness was operative at Seaford College? Was it that pupils happened always to be away from the school houses at times of incidence of the virus? Or was the physical scale of patches of the virus larger than the environs of the school? Unlike the single sharp attack at St. John's, Leatherhead, there was a long extended incidence of new cases at Seaford College. This would make it improbable that pupils always happened to be away from the school houses, and so we think this was a case in which patches of the virus had a scale greater than the horizontal spacing of the school houses.

King Edward's, Witley

The data from King Edward's leads to the following analysis:

$N = 314$, $M = 236$, $p = 0.75$

House	n	np	Actual no. of cases	Fluctuation $\div \sqrt{np(1-p)}$
Ridley	42	31.57	37	1.93
St. Bridget's	38	28.56	28	0.21
Wakefield	43	32.32	36	1.29
Elizabeth	35	26.31	30	1.44
Edward	42	31.57	27	1.63
Tudor	40	30.06	30	0.02
Grafton	42	31.57	25	2.34
Queen's	32	24.05	23	0.43

As with Queen Margaret's, York, King Edward's comes close to the point at which non-randomness in the distribution of cases into houses is to be suspected.

The geographical placing of the houses in pairs did not lead to meaningful correlations, however, as it did in our investigation at Howell's School in Chapter 5 of this volume.

Schools in which a Non-random Distribution of Cases with Respect to Houses is Essentially Certain

First, we give a number of schools for which normal statistical criteria lead to the judgement that non-random processes affected the distribution of influenza victims into houses. Then we shall pass to still more overwhelming examples.

Shiplake, Henley-on-Thames
$N = 266$, $M = 77$, $p = 0.29$

House	n	np	Actual no. of cases	Fluctuation $\div \sqrt{np(1-p)}$
A	57	16.50	25	2.48
B	68	19.68	18	0.45
C	73	21.13	12	2.36
D	68	19.68	22	0.62

Sutton Valence, Maidstone
$N = 260$, $M = 200$, $p = 0.77$

House	n	np	Actual no. of cases	Fluctuation $\div \sqrt{np(1-p)}$
A	63	48.46	36.88	3.46
B	61	46.92	56.48	2.91
C	64	49.23	53.16	1.17
D	59	45.38	53.49	2.51

Stowe, Buckingham
$N = 660$, $M = 327$, $p = 0.50$

House	n	np	Actual no. of cases	Fluctuation $\div \sqrt{np(1-p)}$
Chatham	66	32.70	38	1.31
Bruce	65	32.20	45	3.19
Temple	72	35.67	37	0.31
Grenville	73	36.17	26	2.39
Cobham	62	30.72	37	1.63
Grafton	70	34.68	25	2.32
Nugent	15	7.43	3	2.30
Chandos	72	35.67	35	0.16
Lyttleton	64	31.71	31	0.18
Stanhope	33	16.35	7	3.27
Walpole	68	33.69	43	2.27

St. Audrie's School, West Quantoxhead
$N = 220$, $M = 67$, $p = 0.30$

House	n	np	Actual no. of cases	Fluctuation $\div \sqrt{np(1-p)}$
Main	139	41.7	26	2.91
Olympus	9	2.7	3	0.22
Orchard	18	5.4	11	2.89
Perigrine	54	16.2	27	3.20

Berkhamsted School, Berkhamsted
$N = 145$, $M = 72$, $p = 0.50$

House	n	np	Actual no. of cases	Fluctuation $\div \sqrt{np(1-p)}$
School	50	25	30	1.41
Incents	49	24.5	14	3.00
St. John's	46	23	28	1.47

Ursuline Convent, Westgate-on-Sea
$N = 114$, $M = 67$, $p = 0.59$

House	n	np	Actual no. of cases	Fluctuation $\div \sqrt{np(1-p)}$
Middle	17	9.99	17	3.47
Hatton	38	22.33	25	0.88
Convent	59	34.68	25	2.57

While these schools would plausibly establish the existence of non-random distributions of influenza cases among houses, and hence in our view establish the patchy incidence of the virus on the scale of separation of school buildings, we now add three further schools whose experiences establish the occurrence of non-random distribution outside all possibility of error.

Eton College, Windsor
$N = 1248$, $M = 441$, $p = 0.35$

House	n	np	Actual no. of cases	Fluctuation $\div \sqrt{np(1-p)}$
COLL	about 70	24.74	1	5.94
DJSG	about 50	17.67	16	0.49
JWR	about 50	17.67	22	1.28
RPCF	about 50	17.67	11	1.97
JF	about 50	17.67	22	1.28
RJGP	about 50	17.67	18	0.10
FPEG	about 50	17.67	15	0.79
JSBP	about 50	17.67	9	2.56
ACG	about 50	17.67	18	0.10
NAR	about 50	17.67	11	1.97
RHP	about 50	17.67	6	3.45
DSS	about 50	17.67	11	1.97
MAN	about 50	17.67	32	4.24
TLH	about 50	17.67	23	1.58

JWT	about 50	17.67	26	2.46
KRS	about 50	17.67	9	2.56
DNC	about 50	17.67	14	1.09
MFW	about 50	17.67	31	3.94
AGR	about 50	17.67	15	0.79
CNCA	about 50	17.67	21	0.99
RHH	about 50	17.67	19	0.39
CAI	about 50	17.67	6	3.45
GDR	about 50	17.67	12	1.68
TSBC	about 50	17.67	13	1.38
DH	about 50	17.67	14	1.09

It will be seen that the fluctuation between the actual number of cases and the average np exceeds $3\sqrt{np(1-p)}$ (i.e. more than 3 standard deviations) for as many as 5 of the houses, and even exceeds $5\sqrt{np(1-p)}$ for COLL. Since a truly vast number of random allotments of 441 cases into the 25 houses would be needed before an allotment with fluctuations as large as these occurred, we can safely conclude that the distribution of influenza victims into houses was not random at Eton.

Several other important conclusions follow. Person-to-person transmission of the virus did not occur during school classes, otherwise COLL could not have escaped essentially unscathed from the epidemic. Nor did person-to-person transmission occur in dormitories, for the reason that all pupils have single bedsitting rooms.

We see no possibility of understanding these results except in terms of a patchy incidence of the influenza virus on a physical scale comparable with, or less than, the spacing of the houses. We suspect that patchiness on such a fine scale must be self-generated in some way.

The figure which Dr. Briscoe kindly attached to our questionnaire contains many remarkable features. The houses commonly show sudden bursts of cases, bursts that are apparently uncorrelated in time from one house to another. Bursts sometimes occur comparatively early in the epidemic, as in MAN, MFW, and TLH. Sometimes, however, bursts come at the end of the epidemic, as in ACG and NAR, while in other houses bursts occur more or less midway. There is evidently no consistent pattern.

From Dr. Briscoe's figure our feeling is that the virus may have entered the houses during the first week of February, and that it contrived to stay around within the houses until mid-March, being 'stirred-up' in some fashion from time to time. The influenza virus as it is shed from a person has but a poor measure of persistence, because shedded virus is essentially free and is quickly exposed to external attack. On the other hand, if the influenza virus falls from space, it would need to be encased in a protective matrix in order that it should survive passage down through the Earth's atmosphere. Such a protective matrix might well permit the virus to continue circulating for a considerable period inside a more or less closed building, to be destroyed eventually in some way, by contact with hot surfaces for example. Dr. Briscoe's figure suggests this possibility somewhat insistently, with the late cases that developed in mid-March coming from a final contact with the last remaining pockets of the virus.

St. Mary's, Calne

We come now to another remarkable example of the relation of houses and school buildings to the incidence of the influenza virus. The data for St. Mary's lead to the following table:

$N = 235$, $M = 72$, $p = 0.31$

House	n	np	Actual no. of cases	Fluctuation $\div \sqrt{np(1-p)}$
School	55	16.85	35	5.31
St. Prisca's	25	7.66	12	1.88
St. Cecilia's	48	14.71	15	0.09
New House	57	17.46	2	4.44
Study Bedrooms	8	2.45	3	0.42
Penthouse	8	2.45	1	1.11
Corridor	8	2.45	1	1.11
Mews	27	8.27	3	2.20

If one were considering the experience at St. Mary's in isolation from all other schools, it might seem tempting to ascribe the significantly low incidence of cases for New House to the single rooms enjoyed by the pupils who occupy that building. But single rooms for *all* pupils at Eton College did not prevent an attack-rate of $p = 0.38$, higher than for St. Mary's. Moreover, this explanation would be internally self-contradictory on two further counts. First, quite simply, if there were person-to-person transmission in dormitories with several beds, there would also be person-to-person transmission in school classes, from which the occupants of New House could not have escaped. Second, the dormitory diagrams kindly supplied by St. Mary's show that there was no person-to-person transmission in dormitories with several beds.

These dormitory diagrams can be analysed in the way that we used in Chapter 5 for the dormitories at Howell's. For this we assign 35 cases to School House, 12 to St. Prisca's, and 15 to St. Cecilia's. We can then calculate how on the average these assigned numbers would fall randomly into the dormitories, and can then compare with the actual incidence of cases among the dormitories. For the purpose of setting out a table of results, we classify dormitories according to the number of beds which they contain:

No. of beds in dormitory	*No. of such dormitories in School House, St. Prisca's and St. Cecilia's*	*Average no. expected in such dormitories if assigned cases are distributed randomly*	*Actual no. of cases*
13	1	8.27	9
12	1	7.67	7
8	2	8.93	9
7	2	4.38	6
6	2	7.67	7
5	4	7.87	7

4	2	2.50	4
3	7	9.04	7
2	5	4.47	4
1	2	1.27	2

The fluctuations of actual numbers from the random averages are small, showing that the dormitory distribution of cases was essentially random–there was no detectable person-to-person transmission in dormitories, the same situation as at Howell's.

We conclude that while School House was prone to the incidence of influenza cases, New House was particularly resistant, a statement that we can make at a very high confidence level. From the School Sister we understand that, although detailed records are not now available of an influenza epidemic which occurred at St. Mary's three years ago, the resistant pattern of New House was also maintained on that occasion.

Oakham School, Rutland

An attack pattern rather similar to that at St. Mary's, Calne, occurred at Oakham School. The data for this school leads to the following table:

$N = 583$, $M = 282$, $p = 0.48$

House	n	np	Actual no. of cases	Fluctuation $\div \sqrt{np(1-p)}$
School	60	28.80	7	5.63
Buchanans	47	22.56	19	1.04
Wharflands	80	38.40	61	5.06
Chapmans	82	39.36	25	3.17
Deanscroft	76	36.48	36	0.11
RH	77	36.96	57	4.58
Hodge Wing	38	18.24	15	1.05
Peterborough	60	28.80	30	0.31
Lincoln	63	30.24	32	0.44

Here School House was extraordinarily free from infection, while Wharflands and RH were exceptionally exposed to it, a statement that we can make with a very high degree of confidence.

Schools Showing Dramatic Pulses

Headington, Oxford

The return to our questionnaire from Headington gives absences of boarders from classes, and it does so with respect to houses. In all, there were 356 day-absences from classes by boarders. The return also shows that a total of 109 boarders were victims of influenza, so that the average number of day-absences per influenza case among boarders was 356/109 = 3.27. Assuming the same number of day-absences per boarder for each house separately, the number of influenza cases for each house takes the value given in the fourth column of the following table:

$N = 229$, $M = 109$, $p = 0.48$

House	n	np	Actual no. of cases	Fluctuation $\div \sqrt{np(1-p)}$
Davenport	34	16.18	9	2.47
Hillstow	63	29.99	23	1.77
Latimer	46	21.90	44	6.52
Napier	42	19.99	18	0.62
Celia-Marsh	44	20.94	14	2.10

It hardly needs this analysis, however, to tell us that the situation for Latimer was far from being a mere random fluctuation. The outburst of absences for Latimer over the week beginning 6th February shows that this particular house suffered a sharp, severe pulse of the virus, which did not hit either the other four houses or the day pupils. The pulse must have occurred around 3rd February.

Giggleswick, Settle
Stoke College, Stoke-by-Clare

The returns from these schools show very steep rises in class absences, implying the incidence of high concentrations of the virus in patches sufficient in size to envelop the whole school, rather than only a single house. These sudden pulses for Giggleswick and Stoke College may be contrasted with the broad, long-continued incidence of the virus experienced by King Edward's, Witley. This marked difference in modes of incidence of the virus corresponds very well with the situation that we found ourselves among west country schools.

Upholland College, Skelmersdale

The return from Upholland College gives new cases rather than absences. The rise is once again extremely steep. Moreover, the incidence of the virus was maintained for at least two weeks, leading in the end to an exceedingly high total attack-rate, $p = 0.92$.

In our earlier report (Chapter 5) we remarked that the number of cases sometimes rises much too steeply to be explained in terms of person-to-person transmission. Upholland College, Stoke College, and Giggleswick are examples of such rises. They are caused in our view by sudden pulses in the incidence of the virus, on a horizontal length scale larger than the exceedingly local pulse which hit Latimer House at Headington, Oxford.

What may well have been the most devasting attack of all was reported from Queenswood, Hatfield. The Headmistress writes that the attack was so extreme as to make the keeping of class records seem a thing of little moment:

"I can only say that in the first 48 hours (1st February–2nd February) 212 girls in the age range 11–18 years were affected and a further 150 girls in the days between then and 7th February."

In contrast, the Headmistress of St. Michael's, Oxted, writes:

"I am delighted to say that we had no epidemic of influenza. This school very rarely, if ever, has epidemics. I think we are too high on a hill."

We conclude this report by referring to a most extraordinary situation which occurred at Rydal School, Colwyn Bay. The senior school, comprised of 9 boarding houses, is scattered over a one square mile area of a north-facing hill slope, whereas the preparatory school is right at the top of the hill, buried in woodland. Details for boarders in the senior and preparatory sections taken separately, (including two waves of influenza separated by approximately 4 weeks), are given below:

Rydal School, Colwyn Bay (Senior School)

$N = 289$, $M = 109$, $p = 0.48$

House	n	np	Actual no. of cases	Fluctuation $\div \sqrt{np(1-p)}$
A	24	9.12	13	1.63
B	46	17.48	14	1.06
C	32	12.16	13	0.31
D	37	14.06	15	0.32
E	30	11.40	8	1.28
F	26	9.88	11	0.45
G	42	15.96	15	0.30
H	41	15.58	11	1.47
I	11	4.18	9	2.99

Rydal School, Colwyn Bay (Preparatory School)

$N = 121$, $M = 0$, $p = 0.0$

With the possible exception of House I, the distribution of cases in the secondary school is random with respect to houses. Taking this feature together with the rather protracted period of illness at Rydal, we might conclude that patches of the incident virus were large compared with overall spread of the houses, about one mile. How then, one may ask, were pupils housed in the preparatory school on the hill top completely unscathed? Our provisional answer must be that the hill-top location, as in the case of St. Michael's, Oxted, and perhaps the woodland around the school, had a part to play in its exceptional resistance.

8

SCHOOLS INFLUENZA SURVEY 1985

(by F.H., N.C.W. and J.W.)

In April 1985 we repeated our earlier influenza survey by sending out questionnaires to a sample of independent schools. From a nationwide sample of nearly 300 schools that were written to we received responses from about 250. The patterns of incidence were in most important respects identical to our findings from the 1977/78 survey when the HINI subtype seemed to be the dominant strain. Whereas our earlier survey revealed epidemic peaks clustering near the first week of February 1978, our later survey showed peaks of incidence extending well into March, an effect that we attribute to the abnormal meteorological patterns that prevailed over the British Isles during this winter and spring. Another notable difference was the protracted nature of many of the 1985 epidemics, and also a lower average attack rate as compared with the 1977/78 season.

Below is a synopsis of the results for the boarding schools in our sample arranged as before according to our criterion of randomness of incidence amongst houses. In those instances where only absence data were available we assumed that the average number of days absent per case is about 3 days for all the houses within the school.

Schools with Possibly Random Distribution of Influenza Cases into Houses

Queen's College, Taunton

Data from Queen's College were given for two houses of residence, one house for Junior pupils the other for Senior pupils. The peak of the epidemic was on 15 March 1985. The outbreak appears to have been somewhat protracted in time, extending over 3 weeks. Our analysis of distribution with respect to the two houses is as follows:

$N = 292$, $M = 57$, $p = M/N = 0.20$

House	n	np	Actual no. of cases	Fluctuation $\div \sqrt{np(1-p)}$
Junior	75	14.64	16	0.40
Senior	217	46.36	41	-0.23

In this school it would seem that the virus came down in a patch (or patches) that were large compared with the separation of the two houses, so that both houses were more or less evenly affected.

Cranborne Chase School, Wiltshire

For this school data was available both with respect to classes as well as with respect to houses of residence. For the house data our analysis is thus:

$N = 137$, $M = 49$, $p = 0.36$

House	n	np	Actual no. of cases	Fluctuation $\div \sqrt{np(1-p)}$
Cerne	34	12.16	9	-1.13
Iwerne	33	11.80	15	1.16
Tarrant	34	12.16	10	-0.77
Avon	36	12.88	15	0.74

There is no evidence here for an attack that differentiated between the four houses. The classroom data, on the other hand, exhibits marginal evidence for a daytime incidence of the virus that distinguished between classrooms.

Class	n	np	Actual no. of cases	Fluctuation $\div \sqrt{np(1-p)}$
6Alpha	19	6.85	5	-0.88
L6A	19	6.85	10	1.51
5B	20	7.21	11	1.77

4C	20	7.21	3	-1.96
3D	16	5.12	6	0.12
2E1	23	8.29	5	-1.43
1E2	19	6.85	9	1.03

The peak of the Cranborne Chase epidemic was on 7 March and the duration was about 2 weeks.

Caterham School, Surrey

Data for this school showed a protracted epidemic lasting for over 3 weeks. The epidemic peaked on 20 March. The attacks on the 4 houses were only marginally non-random as the following analysis shows:

$N = 201$, $M = 32$, $p = 0.16$

House	n	np	Actual no. of cases	Fluctuation $\div \sqrt{np(1-p)}$
J.	61	9.71	15	1.85
V.	48	7.64	9	0.54
T.	47	7.48	4	-1.39
B.	45	7.16	4	-1.29

It is possible that individual viral patches which fell over the school were small compared to the separation of houses, but a prolonged period of incidence erased all but a marginal preference for house J.

Headington School, Oxford

Data for Headington show a generally similar pattern to Caterham, with possibly a similar inference. Again the epidemic was spread over 3 weeks (1-21 March) with an attack rate of 16 per cent. The analysis below shows the distribution according to houses.

$N = 219$, $M = 34$, $p = 0.16$

House	n	np	Actual no. of cases	Fluctuation ÷ $\sqrt{np(1-p)}$
Celia Marsh	42	6.52	3	-1.50
Davenport	33	5.12	9	1.86
Hillstow	60	9.32	8	-0.47
Latimer	43	6.68	8	0.56
Napier	41	6.37	6	-0.16

The situation here contrasts markedly with our earlier analysis of 1978, where Latimer house showed a fluctuation of 6.52 times the standard deviation. The contrast also shows up in the sharpness of the 1978 attack compared with the protracted nature of the attack in 1985.

Schools in which a Non-random Distribution of Cases with Respect to Houses is Possibly Established

As in our analysis of the 1977/78 epidemic we now give cases for which normal statistical criteria lead to the conclusion that non-random processes affected the distribution of influenza victims into houses.

Mill Hill School, London

$N = 282$, $M = 47$, $p = 0.17$

House	n	np	Actual no. of cases	Fluctuation ÷ $\sqrt{np(1-p)}$
A	58	9.67	10	0.12
B	52	8.67	6	-0.99
C	60	10.00	17	2.42
D	64	10.67	11	0.11
E	48	8.00	3	-1.94

SCHOOLS INFLUENZA SURVEY 1985

This school had an epidemic lasting 2 weeks and peaking at 4 March. The above analysis shows that it just qualifies for entry into the non-random category.

St. John's School, Leatherhead

A similar situation persisted for St. John's, Leatherhead as the following analysis shows:

$N = 200$, $M = 36$, $p = 0.18$

House	n	np	Actual no. of cases	Fluctuation $\div \sqrt{np(1-p)}$
South	57	10.26	13	0.94
West	54	9.72	15	1.87
East	56	10.08	4	-2.11
Churchill	33	5.94	4	-0.88

The epidemic here was somewhat protracted (lasting over 2 weeks) with a peak date on 26 January.

Queen Margaret's School, York

The epidemic peaked on 20 February, with a total duration of two weeks.

$N = 198$, $M = 66$, $p = 0.33$

House	n	np	Actual no. of cases	Fluctuation $\div \sqrt{np(1-p)}$
Red	42	14.00	11	-0.94
Howard	29	9.67	13	1.35
Scarborough	26	8.67	10	0.59
School	27	9.00	11	0.86
West Wing	29	9.67	17	2.93
Study Wing	28	9.33	2	-2.91
Carr	17	5.67	2	-1.86

Schools in which a Non-random Distribution with Respect to Houses is Essentially Certain

We now pass on to our third category where by normal statistical criteria a non-random distribution of victims by houses cannot be denied.

Howell's School, Llandaff

$N=62$, $M=31$, $p=0.50$

House	n	np	Actual no. of cases	Fluctuation $\div \sqrt{np(1-p)}$
Oaklands	27	13.52	23	3.66
Taylor	21	10.5	5	-2.40
Bryn Taff	14	7.0	3	-2.14

The duration of the epidemic was less than 2 weeks with a peak occuring in the week 31 January - 7 Feburary 1985.

Woodbridge School, Woodbridge

$N=120$, $M=39$, $p=0.325$

House	n	np	Actual no. of cases	Fluctuation $\div \sqrt{np(1-p)}$
School	54	17.55	34	4.78
Tallent's	23	7.48	5	-1.10
Queen's	29	9.43	0	-3.74
H.M.'s	14	4.55	0	-2.60

The epidemic lasted for less than 2 weeks and peaked on 20 February.

SCHOOLS INFLUENZA SURVEY 1985

Howell's School, Debigh, Clwyd

$N = 225$, $M = 96$, $p = 0.43$

House	n	np	Actual no. of cases	Fluctuation $\div \sqrt{np(1-p)}$
B	30	12.80	17	1.55
P	35	14.93	26	3.78
A	29	12.37	9	-1.27
G	33	14.08	18	1.38
M	26	11.09	11	-0.04
S	72	30.72	15	-3.75

The epidemic was sharply spiked in time (lasting less than 1 week) with a peak occuring on 19 February 1985.

King Edward's School, Witley

$N = 412$, $M = 150$, $p = 0.364$

House	n	np	Actual no. of cases	Fluctuation $\div \sqrt{np(1-p)}$
Ridley(B)	32	11.65	25	4.90
Wakefield(B)	36	13.11	8	-1.77
Edward(B)	36	13.11	24	3.77
Gaften(B)	36	13.11	10	-1.08
St.Bridget's(G)	39	14.20	17	0.93
Elizabeth(G)	42	15.29	21	1.83
Tudor(G)	41	14.93	11	-1.27
Queen's(G)	41	14.93	11	-1.27
*Queen Mary(B)	49	17.84	6	-3.52
*Copeland(G)	60	21.84	17	-1.30

*stands for Junior house; rest are Senior houses; B:Boys; G:Girls

The epidemic lasted less than 2 weeks with a peak occurring on 22 February 1985.

EPIDEMICS SHOWING DRAMATIC NON-RANDOMNESS AT HARROW AND ETON

We conclude our present account with an analysis of data for the two most famous independent schools in England. Each school showed dramatic evidence of non-randomness in the distribution of victims with respect to school houses. The Harrow epidemic lasted a little over a week and peaked on 22 March. The Eton epidemic was much more protracted and lasted for nearly 4 weeks, with the main peak occurring at the end of March. The data for Eton suggests that a patchy distribution of viral particles descended over the school at some particular moment of time during the night hours, and in so descending established preferences for certain houses, notably for house no. 24. The virus continued thereafter to circulate within the houses for nearly 4 weeks.

Harrow School, Harrow

$N = 756$, $M = 28$, $p = 0.037$

House	n	np	Actual no. of cases	Fluctuation $\div \sqrt{np(1-p)}$
Brad.	63	2.33	2	-0.22
Dru.	60	2.22	1	-0.84
Ess.	69	2.56	0	-1.63
Gro.	89	3.30	15	6.57
Hea.	74	2.74	0	-1.69
Kno.	65	2.41	0	-1.58
Mor.	66	2.44	0	-1.59
New.	74	2.74	8	3.24
Par.	67	2.48	0	-1.61
Ren.	59	2.19	0	-1.51
W.A.	70	2.59	2	-0.38

SCHOOLS INFLUENZA SURVEY 1985

Eton College, Windsor

$N = 1270$, $M = 226$, $p = 0.18$

House	n	np	Actual no. of cases	Fluctuation $\div \sqrt{np(1-p)}$
1	about 70	12.46	22	2.98
2	about 50	8.90	9	0.04
3	about 50	8.90	4	-1.81
4	about 50	8.90	4	-1.81
5	about 50	8.90	9*	0.04
6	about 50	8.90	3	-2.18
7	about 50	8.90	14	1.89
8	about 50	8.90	10	0.41
9	about 50	8.90	0	-3.29
10	about 50	8.90	10	0.41
11	about 50	8.90	15	2.26
12	about 50	8.90	1	-2.92
13	about 50	8.90	6	-1.07
14	about 50	8.90	6	-1.07
15	about 50	8.90	7	-0.70
16	about 50	8.90	9*	0.04
17	about 50	8.90	4	-1.81
18	about 50	8.90	19	3.74
19	about 50	8.90	5	-1.44
20	about 50	8.90	6	-1.07
21	about 50	8.90	10	0.41
22	about 50	8.90	10	0.41
23	about 50	8.90	6	-1.07
24	about 50	8.90	29	7.43
25	about 50	8.90	8	-0.33

* We have used an estimate of 9 cases where no data was available.

Index

Al Mufti, S., 18
Alps, effect on viral input, 24
Antigenic drift, 61
Antigenic shift, 61
Arrhenius, S., 18
Asian flu pandemic of 1957–58, 32
Athens, Plague of, 47 ff, 85
Atlantic College, 68 ff
Attack rates of influenza etc.
    Boarders *vs* Day pupils, 12, 13, 14
    Histogram of attack rates in schools, 2
    Household attack patterns, 11, 12
Attack rates of whooping cough, measles and infective jaundice, 7, 8
Australia, influenza in, 30

Balfour House, 69
Berkhamstead School, 88, 95
Black Death, 24, 25
Briscoe, Dr. J., 14
Britten, R.A., 84
Brownlee, D.E., 84

Cardiff, City of, 7, 8
Cardiff High School, 69
Caterham School, Surrey, 107
Cirencester, influenza rates in, 23
Claus, G., 84
Clifton College, 69
Colombo, Sri Lanka, 40 ff
Comets
    and diseases, 16 ff, 79 ff
    and life, 81 ff
    Comet Kohoutek, 81, 82
    Halley's comet, 16
    orbits of comets, 9
Cometary debris in the stratosphere, 84
Conference epidemic in Sri Lanka, 40 ff
Cranborne Chase School, Wiltshire, 106
Creighton, Charles, 1, 11, 65, 76

INDEX

Descent of small particles through the atmosphere, 21 ff
Details of vertical incidence of viruses etc., 27 ff
Diatoms, 19 ff
Dispersal of viruses at ground level, 33-35

Effect of cold weather on viral incidence, 34
Endogenous viruses, 60
Eton College, Windsor, 14, 28, 33, 88, 96-98, 113
Evidence against the horizontal transmission of certain viruses, 7 ff
Evolution, 63

Farrar, I.H., 84

Genes in pieces, 59
Genetic recombination, 61
Genetic viruses, 57 ff
Giggleswick School, Settle, 88, 102
Glamorgan, Vale of, 7, 8
Global infall patterns of influenza, 30
Growth cycle of plants and summer epidemics, 35

H1 N2 influenza outbreak of 1977-78, 1, 2, 57, 62, 67 ff, 87 ff
H. Influenzae, 1
Haemoglobin in legumes, 60
Haldane, J.B.S., 79
Halley's Comet, 16
Hardtack atmosphere nuclear bomb (1958), 26, 27
Harrow School, 112
Headington School, Oxford, 88, 101, 107, 108
Himalayas, effect on particle infall, 23, 26
Histogram of influenza attack rates in schools, 2
Hodge, P.W., 84
Holladay, A.J., 54
Hoover, M.J., 18
Hoover, R.B., 18
Hope-Simpson, R.E., 11, 23, 24, 30
Household attack patterns of influenza, 11, 12
Howells, C.W.L., 78
Howell's Girls School, Llandaff, 68 ff, 110
Howell's School, Denbigh, Clwyd, 111
Hoyle, F., 18

Influenza
    Attack rates: boarders *vs* day pupils, 12-14
    Attack rate in schools, 2
    Household attack patterns, 11, 12
    in Australia, 30
    in Cirencester, 23
    in Prague, 23

INDEX

    in Sri Lanka, 30
    in Sweden, 30
    Survey in Schools, 87 ff, 105 ff
Infective Jaundice, 7, 8
Input of zodiacal dust onto Earth, 4
Irvine, W.M., 57

Jenkins, Dr. P., 7
Jowett, B., 47 ff

Kalkstein, M.I., 26
Kaplan, M.W., 65
King Edward's School, Witley, 88, 93, 94, 111
Kohoutek, Comet, 81, 82
Kushner, D.J., 18

Llanishen High School, 69

Mann, P.G., 11, 78
Mantle, J., 67
Measles, 4, 7
Melbourne, Influenza in, 30
Microfossils in meteorites, 84
Miller, Stanley, 79
Mill Hill School, London, 108, 109
Munro, H.A.J., 47

Nagy, B. 84

Oakham School, Rutland, 88, 100
Oparin, A.I., 79
Oparin-Haldane theory, 79 ff
Orbits of short period comets, 9
Organisms in Apollo II spacecraft, 20
Osler, 49 ff
Ozone as indicator of stratospheric air movements, 25

Page, D.L., 51
Palmer, Dr. H., 51, 52
Panspermia theory, 18, 21
Pasteur, L., 1
Peloponnesian Wars, 44 ff
Peradeniya, University of, 41
        Botanic Gardens, 45
Pereira, M.S., 78
Pertussis, 5-7
Pfeiffer, R., 1
Philip, R.E., 11
Plague of Athens, 47 ff, 85
Ponnamperuma, Cyril, 79

Poole, J.C.F., 54
Prague, influenza rates in, 23
Prebiotic molecules, 79 ff
Primeval soup, 79 ff

Queen Margaret's School, York, 88, 91, 92, 109
Queen's College, Taunton, 105, 106
Queenswood, Hatfield, 88

Radiation resistance of bacteria and viruses, 18 ff
Rajan, R.S., 84
Random genetic drift, 61
Respiratory Syncytial Virus (RS), 26, 28, 30
Rh–102 as a tracer of viral transport, 26, 27
Rydal School, Colwyn Bay, 88, 103

Schools Influenza Survey, 87 ff, 105 ff
Seaford College, Petworth, 88, 92, 93
Sekanina, Z., 17
Selkon, J.B., 57
Shiplake, Henley-on-Thames, 88, 94
Shrewsbury, J.F.D., 51
Size of microorganism for safe entry, 21
Small pox, 4, 47 ff
Sri Lanka
    influenza data, 30
    conference epidemic, 40 ff
Statistical considerations relating to influenza, 89, 90
St. Audrie's School, West Quantoxhead, 88, 95
St. Cyre's School, 69
St. John's, Letherhead, 88, 90, 91, 109
St. Mary's Calne, 88, 98-100
St. Michael's, Oxted, 88, 104
Stoke College, Stoke-by-Clare, 88, 102
Stowe, Buckingham, 88, 95
Sutton Valence, Maidstone, 88, 94
Sweden, influenza data, 30

Thucydides, 47 ff, 85
Tomandle, D., 84
Trees and the dispersal of viruses, 35
Tyrell, D.A.J., 76

University of Peradeniya, 41
Upholland College, Skelmersdale, 88, 102
U-pond, 19
Ursuline Convent, Westgate-on-Sea, 88, 96

INDEX

Vale of Glamorgan, 7,8
Vertical incidence theory, 16 ff, 27 ff
Viroids, 58 ff
Viroids as addresses, 59, 60

Watkins, Prof. John, 57
Webster, R.G., 65
Whooping cough, 5-7
Wickramasinghe, N.C., 18
Wienstein, L., 66
Woodbridge School, Woodbridge, 110

Z-trench, 19
Zodiacal particles, 4